P9-DZZ-668

# THE 10,000 YEAR EXPLOSION

# THE 10,000 YEAR EXPLOSION

## HOW CIVILIZATION ACCELERATED HUMAN EVOLUTION

**Gregory Cochran** and **Henry Harpending**

A Member of the Perseus Books Group
*New York*

Published by Basic Books,
A Member of the Perseus Books Group



Books published by Basic Books are available at special discounts for bulk purchases in the United States by corporations, institutions, and other organizations. For more information, please contact the Special Markets Department at the Perseus Books Group, 2300 Chestnut Street, Suite 200, Philadelphia, PA 19103, or call (800) 810-4145, ext. 5000, or e-mail special.markets@perseusbooks.com.

Designed by Pauline Brown

Library of Congress Cataloging-in-Publication Data

Cochran, Gregory.
The 10,000 year explosion : how civilization accelerated human evolution / Gregory Cochran and Henry Harpending.
p. cm.
Includes bibliographical references and index.
ISBN 978-0-465-00221-4 (alk. paper)
1. Human evolution. 2. Genetics. I. Harpending, Henry. II. Title. III. Title: Ten thousand year explosion.
[DNLM: 1. Evolution. 2. Hominidae. 3. Civilization. 4. Genome—genetics. GN 281.4 C663z 2009]
GN281.4.C632 2009
599.93'8—dc22

2008036672

10 9 8 7 6 5 4 3 2 1

TO OUR FAMILIES

# CONTENTS

Preface ix

1 OVERVIEW: CONVENTIONAL WISDOM 1

2 THE NEANDERTHAL WITHIN 25

3 AGRICULTURE: THE BIG CHANGE 65

4 CONSEQUENCES OF AGRICULTURE 85

5 GENE FLOW 129

6 EXPANSIONS 155

7 MEDIEVAL EVOLUTION: HOW THE ASHKENAZI JEWS GOT THEIR SMARTS 187

CONCLUSION 225

Notes 229

Glossary 243

Bibliography 253

Credits 267

Index 269

# PREFACE

For most of the last century, the received wisdom in the social sciences has been that human evolution stopped a long time ago—in the most up-to-date version, before modern humans expanded out of Africa some 50,000 years ago. This implies that human minds must be the same everywhere—the "psychic unity of mankind." It would certainly make life simpler if it were true. Unfortunately, a recent halt to evolution also implies that human *bodies* must be the same everywhere, which is obviously false. Clearly, received wisdom is wrong, and human evolution continued. In the light of modern evolutionary theory, it is difficult to imagine how it could have been otherwise.

Since the social sciences—anthropology in particular—haven't exactly covered themselves with glory, we have decided to take a new tack in writing this book, one that takes the implications of evolutionary theory seriously while cheerfully discarding unproven anthropological doctrines. Our approach leans heavily on genetics—and with genetic information accumulating at an incredible rate, due to the ongoing revolution in molecular biology, it is an approach that we believe has been very fruitful. At the same time, we make use of paleontology, archaeology, and good old-fashioned history to support our arguments. We think that it is mistake to neglect any relevant information.

A lot of our work could be called "genetic history." It's a new kind of history: We share the usual facts, but we use a very different explanatory framework. Traditional historians tell the stories of battles and kingdoms and great men. Some study the history of ideas, or of science and technology. Quantitative historians examine commerce and demographic trends. We, however, are interested in the historical factors that have influenced natural selection in humans, particularly those having to do with the creation and spread of new, favorable alleles. This means that when a state hires foreign mercenaries, we are interested in their numbers, their geographic origin, and the extent to which they settled down and mixed with the local population. We don't much care whether they won their battles, as long as they survived and bred. We're not particularly interested in their cultural baggage unless it changes selection pressures or influences gene flow.

Conventional social sciences, such as history and anthropology, chiefly concern themselves with brain software, by which we mean cultural developments such as mores, mythology, or social structure. Genetic history addresses changes in the underlying hardware, changes in body and brain, which also matter. If they didn't, dogs really could play poker.

For an anthropologist, it might be important to look at how farmers in a certain region and time period lived; for us, as genetic historians, the interesting thing is how natural selection allowed agriculture to come about to begin with, and how the new pressures of an agricultural lifestyle allowed changes in the population's genetic makeup to take root and spread. We take this same approach whether we are looking at the human revolution of the Paleolithic period, the agricultural revolution of 10,000 years ago, or the modern development of the Ashkenazi Jews of Europe, who rose to intellectual prominence in the West quite recently after experiencing unique natural-selection pressures during the Middle Ages.

We arrived at this approach via different paths. Gregory Cochran was trained as a physicist—and following theory where it leads, no matter how odd, comes naturally to a physicist. Henry Harpending began his graduate career with enthusiasm for the social sciences. Years of disillusionment with the status quo in these fields led him to into demography and genetics, areas where he believes there is real theory with strong links to the rest of the sciences.

Many colleagues have provided suggestions, ideas, and objections to the material in this book. We are grateful to all of them for their help and their criticism. Particularly prominent among these colleagues are John Hawks of the University of Wisconsin; Robert Moyzis and Eric Wang of the University of California at Irvine (Wang is now at Affymetrix Corporation); and Alan Rogers, Doug Jones, and Renee Pennington of the University of Utah. We have also benefited from discussions with members of the Human Biodiversity Internet discussion group run by Steve Sailer.

We are indebted to Kristen Hawkes, James O'Connell, Dennis O'Rourke, and Jon Seger of the University of Utah; Gregory Clark of the University of California at Davis; Alan Fix of the University of California at Riverside; Montgomery Slatkin of the University of California at Berkeley; Kim Hill of Arizona State University; Bruce Lahn of the University of Chicago; Mel Konner of Emory University; Jeremy Stone of Catalytic

Diplomacy; Razib Khan of the GNXP Web site; James Lee of Harvard University; Rosalind Arden of King's College London; Phil Rushton of the University of Western Ontario; and Balaji Srinavasan of Stanford University. Parts of this work were supported by the Unz Foundation.

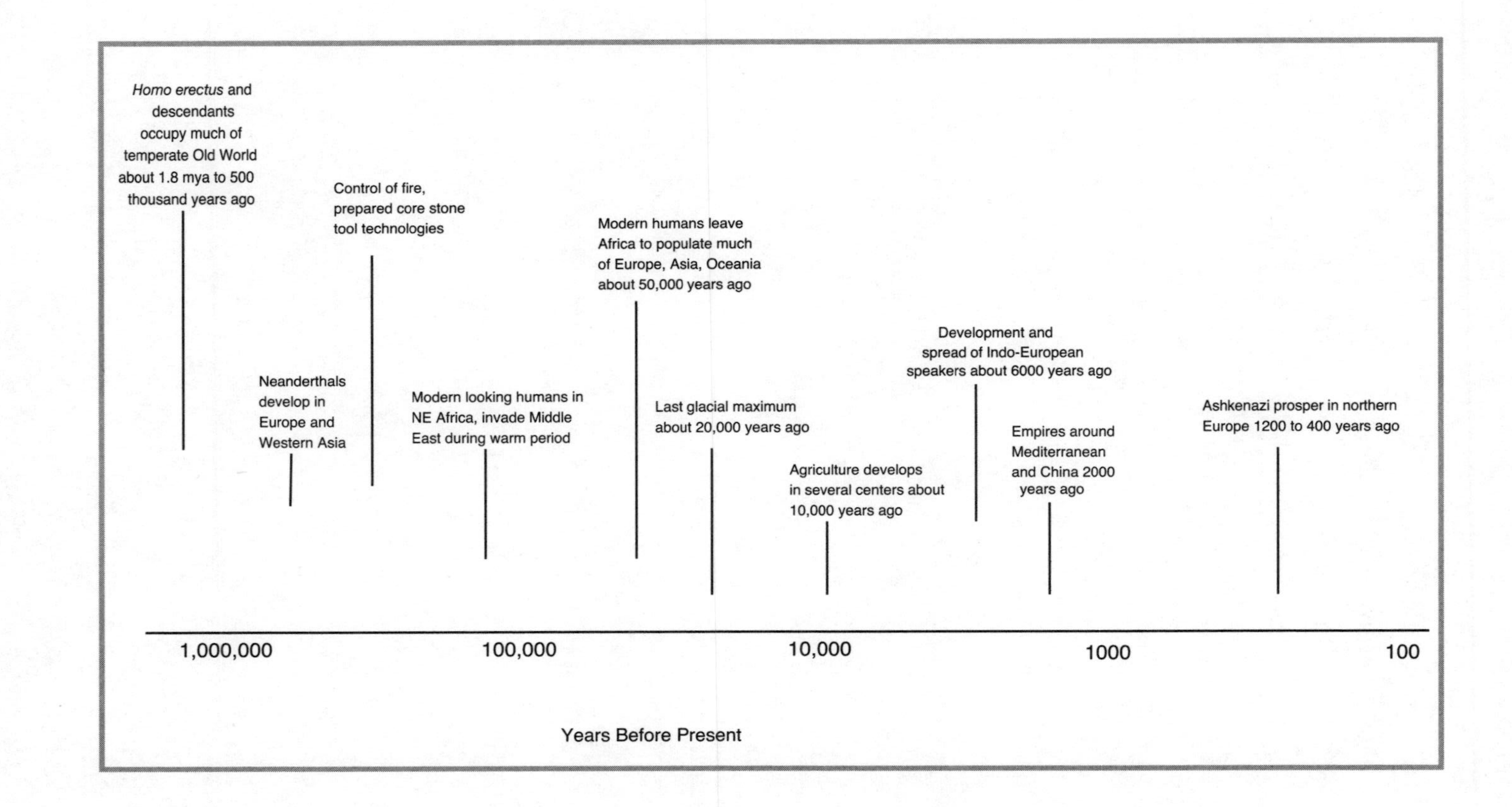

Homo erectus and descendants occupy much of temperate Old World about 1.8 mya to 500 thousand years ago
Neanderthals develop in Europe and Western Asia
Control of fire, prepared core stone tool technologies
Modern looking humans in NE Africa, invade Middle East during warm period
Modern humans leave Africa to populate much of Europe, Asia, Oceania about 50,000 years ago
Last glacial maximum about 20,000 years ago
Agriculture develops in several centers about 10,000 years ago
Development and spread of Indo-European speakers about 6000 years ago
Empires around Mediterranean and China 2000 years ago
Ashkenazi prosper in northern Europe 1200 to 400 years ago
1,000,000
100,000
10,000
1000
100
Years Before Present

# 1 OVERVIEW: CONVENTIONAL WISDOM

There's been no biological change in humans in 40,000 or 50,000 years. Everything we call culture and civilization we've built with the same body and brain.

**—Stephen Jay Gould**

Something must have happened to weaken the selective pressure drastically. We cannot escape the conclusion that man's evolution towards manness suddenly came to a halt.

**—Ernst Mayr**

We intend to make the case that human evolution has accelerated in the past 10,000 years, rather than slowing or stopping, and is now happening about 100 times faster than its long-term average over the 6 million years of our existence. The pace has been so rapid that humans have changed significantly in body and mind over recorded history. Sargon and Imhotep[1] were different from you genetically as well as culturally. This is a radical idea and hard to believe—it's rather like trees growing noticeably as you watch. But as we will show in the following pages, the evidence is there.

Scientists have long believed that the "great leap forward," some 40,000 to 50,000 years ago in Europe, marked the advent of cultural evolution and the end of significant biological evolution in humans. At this time, the theory goes, humans developed culture, as shown by the sophisticated new tools, art, and forms of personal decoration that emerged in the Upper Paleolithic. Culture then freed the human race from the pressures of natural selection: We made clothes rather than growing fur and built better weapons rather than becoming stronger.

The argument that the advent of behavioral modernity somehow froze human evolution is dependent on the notion of a static environment.[2] In other words, if a population—of humans, wolves, crabgrass, you name it—experiences a stable environment for a long time, it will eventually become genetically well matched to that environment. Simple genetic changes then do little to improve individual fitness, because the species is close to an optimum. An economist would say that all the $100 bills have already been picked up off the sidewalk. In that situation, evolution slows to a crawl. That's not to say that a stable species has reached perfection, but that its life strategy is well implemented. For example, hopping may not be as efficient as walking on all fours (four legs good, two legs bad!), but kangaroos are good at hopping; their bodies are well suited to their style of locomotion. The match of the population to its environment can never be exact, since environments fluctuate, but it can be quite close. For example, there are orchids that imitate a bee so closely in appearance and in odor that bees try to mate with them, and so pollinate the orchids. Some creatures that are well suited to their environment, such as horseshoe crabs, have managed to stay much the same for hundreds of millions of years. They're literally older than the hills.

However, modern humans have experienced a storm of change over the past 50,000 years. We left Africa and settled every continent other than Antarctica. We encountered and displaced archaic humans like Neanderthals—and probably picked up genes from them in the process. An ever-accelerating cultural explosion from the Upper Paleolithic to the Neolithic and beyond led to new technologies and new social forms. Indeed, culture itself has been an increasingly important part of the human environment.

Geographic expansion (which placed us in new environments) and cultural innovation both changed the selective pressures humans experienced. The payoff of many traits changed, and so did optimal life strategy. For example, when humans hunted big game 100,000 years ago, they relied on close-in attacks with thrusting spears. Such attacks were highly dangerous and physically taxing, so in those days, hunters had to be heavily muscled and have thick bones. That kind of body had its disadvantages—if nothing else, it required more food—but on the whole, it was the best solution in that situation. But new weapons like the atlatl (a spearthrower) and the bow effectively stored muscle-generated energy, which meant that hunters could kill big game without big biceps and robust skeletons. Once that happened, lightly built people, who were better runners and did not need as much food, became competitively superior. A heavy build was yesterday's solution: expensive, but no longer necessary. The Bushmen of southern Africa lived as hunter-gatherers until very recently, hunting game with bows and poisoned arrows for thousands of years in that region. They are a small, tough, lean people, less than five feet tall. It seems likely that the tools made the man—the bow begat the Bushmen.

With the invention of nets and harpoons, fish became a more important part of the diet in many areas of the world, and metabolic changes that better suited humans to that diet were favored. Close-fitting clothing provided better protection against cold, allowing people to venture farther north. In cool areas, people needed fewer physiological defenses against low temperatures, while in the newly settled colder regions they needed more such defenses, such as shorter arms and legs, higher basal metabolism, and smaller noses. With the advent of new methods of food preparation, such as the use of fire for cooking, teeth began to shrink, and they continued to do so over many generations. Pottery, which allowed storage of liquid foods, accelerated that shrinkage. Complex biological functions tend to slowly deteriorate when they no longer matter, since mutations that interfere with the function no longer reduce reproductive fitness, and you might think that this would explain these dental changes. However, this trend, which we call "relaxed selection," happens too slowly to be the explanation. Instead, the changes in tooth size must have been driven by positive advantages—possibly because small teeth are metabolically cheaper than large ones.

As the complexity of human speech approached modern levels, there must have been selection for changes in hearing (both changes in the ear and in how the brain processes sounds) that allowed better discrimination of speech sounds. Think of the potential advantages in being just a bit better at deciphering a hard-to-understand verbal message than other people: Eavesdropping can be a life-or-death affair. We have evidence of this, since a number of genes affecting the inner ear show signs of recent selection.[3] Such genes are easy to recognize, since radical changes in them cause deafness. Combined with an increased capacity for innovation, complex speech must also have entailed an increase in the capacity for deception—and must have resulted in selective pressures for changes in personality and cognition that helped people resist Paleolithic con men.

There's a common impression that evolutionary change is inherently very slow, so that significant change always takes millions of years. A more detailed look at the fossil record, combined with evidence from contemporary examples of natural selection, makes it clear that natural selection can proceed quite rapidly, and that the past consists of long periods of near-stasis (in populations that were well matched to their environments) interspersed with occasional periods of very rapid change. Those brief periods of rapid change are poorly represented in the fossil record, since fossilization is rare.

Stephen Jay Gould's position that 50,000 or 100,000 years is an "eye blink," far too short a time to see "anything in the way of evolutionary difference," is simply incorrect.[4] We are surrounded by cases in which selection has caused big changes over shorter time spans, often far shorter; everything from the dog at your feet to corn on the cob is the product of recent evolution.

The most accessible examples are the products of domestication. Domesticated animals and plants often look and act very different from their wild ancestors, and in every such case, the changes took place in far less than 100,000 years. For example, dogs were domesticated from wolves around 15,000 years ago; they now come in more varied shapes and sizes than any other mammal.

Gray wolf

Chihuahua and Great Dane

Their behavior has changed as well: Dogs are good at reading human voice and gestures, while wolves can't understand us at all. Male wolves pair-bond with females and put a lot of effort into helping raise their pups, but male dogs—well, call them irresponsible. There have been substantial changes in dogs in just the past couple of centuries: Most of the breeds we know today are no older than that.

In an extreme example, the Russian scientist Dmitri Belyaev succeeded in developing a domesticated fox in only forty years.[5] In each generation he selected for tameness (and only tameness); this eventually resulted in foxes that were friendly and enjoyed human contact, in strong contrast to wild foxes. This strain of tame foxes also changed in other ways: Their coat color lightened, their skulls became rounder, and some of them were born with floppy ears. It seems that some of the genes influencing behavior (tameness in this case) also affect other traits—so when Belyaev selected for tameness, he automatically got changes in those other traits as well. Many of these changes have occurred as side effects of domestication in a number of species—possibly including humans, as we shall see.

Changes in domesticated plants can be just as impressive. Corn, or maize, which is derived from a wild grass named teosinte, has changed wildly in only 7,000 years. It's hard to believe that maize and teosinte are closely related.

Such dramatic responses to selection aren't isolated cases—they've occurred in many domesticated species and continue to occur today. Evolutionary genetics predicts that substantial change in almost any trait is possible in a few tens of generations, and those predictions are confirmed every day. Selection is used routinely in many kinds of agriculture, and it works: It grows more corn, lots more. You can't argue with corn.

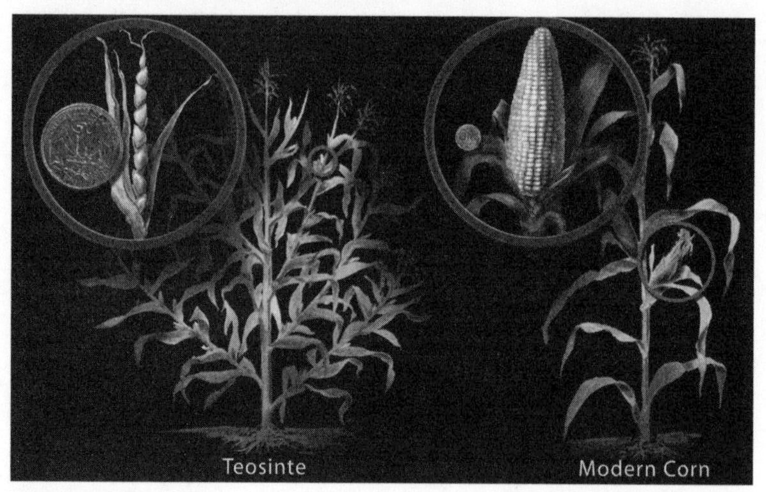


Teosinte and corn

That doesn't keep some people from trying, though. One argument is that domesticated animals and plants are examples of artificial selection and so not relevant. But the process in which some gene variants are favored and gradually increase in frequency is the essence of evolutionary change for both natural and artificial selection. There is no fundamental distinction in the process, just a difference in scale. Furthermore, we have on record examples of entirely natural adaptive change over a few thousand years—the time since the end of the Ice Age.

## AFTER THE ICE

The Ice Age ended (or paused, at any rate) some 11,500 years ago. That caused dramatic environmental changes in many parts of the world, especially in the Northern Hemisphere. The American Southwest turned warmer and drier, becoming the desert it is today, and as it did, the creosote bush appeared there.

Originally from Argentina, its seeds were probably transported north by migratory birds. It thrived in the desert, thanks to its resin-coated leaves; dense lateral roots, which starve out competing plants; and taproot, which can grow up to 15 feet deep into the earth. A number of insects now live in and on creosote bushes: Some have become so specialized that they can eat nothing else. The creosote bush walking-stick looks just like the creosote stems, while a grasshopper, which even has silvery patches that match the shine of the plant's resin, mimics the leaves. All of these creosote-specialized insects in the Southwest have North American ancestors, not South American—and so all their specializations have come into existence in the past 10,000 years.[6]

The end of the Ice Age also brought about a global rise in sea level. Mile-thick continental ice sheets melted, and the sea level rose hundreds of feet. As the waters rose, some mountains became islands, isolating small groups of various species. These islands were too small to sustain populations of large predators, and in their absence the payoff for being huge disappeared. Instead, small elephants had an advantage over large ones, probably because they required less food and reproduced more rapidly. Over a mere 5,000 years, elephants shrank dramatically, from an original height of 12 feet to as little as 3 feet. It is worth noting that elephant generations are roughly twenty years long, similar to those of humans.

But simply getting smaller is hardly a dramatic example of evolution. Indeed, John Tooby and Leda Cosmides (two of the founders of modern evolutionary psychology) have said that "given the long human generation time, and the fact that agriculture represents less than 1 percent of the evolutionary history of the genus *Homo*, it is unlikely that we have evolved any complex adaptations to an agricultural (or industrial) way of life."[7] A complex adaptation is a characteristic contributing to reproductive fitness that involves the coordinated actions of many genes. This means that humans could not have evolved wings, a third eye, or any new and truly complicated adaptive behavior in that time frame. Tooby and Cosmides have argued elsewhere that, therefore, deep mental differences between human populations cannot exist.[8]

We think that this argument concerning the evolution of new complex adaptations is correct, but it underestimates the importance of simple adaptations, those that involve changes in one or a few genes. The conclusion that all humans are effectively the same is unwarranted. We will see not only that we have been evolving at a rapid rate, but that evolution has taken a different course in different populations. Over time, we have become more and more unlike one another as differences among populations have accumulated.

## DOGS

Let's look at dogs again, as they are well understood examples of rapid evolution. They've been domesticated for roughly as long as humans have farmed, and in that time they have changed a great deal. You can see that dog behaviors are derived from the behavioral adaptations of wolves, their ancestral species. There are breeds like the Irish setter that point, and breeds like the Border collie that live to herd other animals. Both breeds show elaborations of behaviors we see in wolves. When wolves first scent a prey, the leading pack members freeze and point rigidly in the direction of the scent. Border-collie herding instinct must also derive from wolf game-herding patterns, but it is greatly exaggerated.

Dogs are much more playful than wolves, and this can probably be understood as retention of juvenile behavior (called "neoteny"). Retaining existing juvenile behavior is accomplished far more easily than evolving a behavior from scratch. Many of the ways in which dogs interact with humans can be understood as a new application of behavioral adaptations designed for a pack—the owner takes on the role of the leader of the pack.

There is no complex behavioral adaptation in dogs without a recognizable precursor in wolves, but that hardly means that all breeds of dogs are the same, or close to it. The testimony of accident statistics is stark: Biting—universal in dogs—is disproportionately distributed among breeds. A survey of U.S. dog attacks from 1982 to 2006 found 1 record of bodily harm attributable to Border collies, but 1,110 records attributable to pit bull terriers.[9]

While there has probably not been enough time for dogs to develop wholly new complex adaptations, there has certainly been enough time to lose some, sometimes in all breeds, but other times only in a subset of dog breeds. Wolf bitches dig birthing dens; a few breeds of dogs still do, but most do not. Wolves go into season in a predictable way, at a fixed time of the year; a few dog breeds do, but most do not. Wolves regurgitate food for weaned cubs, but dogs no longer do so. Male wolves help care for their offspring, but male dogs do not. Any adaptation, whether physical or behavioral, that loses its utility in a new environment can be lost rapidly, especially if it has any noticeable cost. Fish in lightless caves lose their sight over a few thousand years at most—much more rapidly than it took for eyes to evolve in the first place.

In some sense these are evolutionarily shallow changes, mostly involving loss of function or exaggerations and redirections of function. Although changes of this sort will not produce gills or sonar, they can accomplish amazing things. Dogs are all one species, but as we have noted, they vary more in morphology than any other mammal and have developed many odd abilities, including learning abilities: Dog breeds vary greatly in learning speed and capacity. The number of repetitions required to learn a new command can vary by factors of ten or more from one breed to another. The typical Border collie can learn a new command after 5 repetitions and respond correctly 95 percent of the time, whereas a basset hound takes 80–100 repetitions to achieve a 25 percent accuracy rate.

## DIALS AND KNOBS

In the same way, we expect that most of the recent changes in humans are evolutionarily shallow, one mutation deep for the most part. Old adaptations could have been lost in some groups but retained in others. We know of at least one example, which we'll discuss in Chapter 4: Some light-skinned populations, in particular northern Europeans, have lost most of their ability to produce melanin.

Many such changes can be thought of as turning switches or twirling knobs: Biological processes that were once tightly regulated can be turned on all the time, as with lactose tolerance; turned off entirely, as with the caspase 12 gene, which increases the risk of sepsis when intact and which is inactivated in most populations;[10] or turned off selectively, as with the Duffy mutation, a malaria defense that keeps a certain receptor molecule from being expressed on red cells while continuing to be expressed everywhere else. Some other changes are more like turning up the volume (sometimes all the way to eleven), as in some groups that have extra copies of the gene producing amylase, an enzyme present in saliva that aids in digesting starch.[11]

In addition, some behaviors may be the manifestations of genetically influenced alternative behavioral strategies, such as those in a hawk-dove game, as we discuss in Chapter 3: Recent natural selection might have eliminated a particular strategy in some populations or might have materially changed the frequency of existing strategies. Such strategies probably exist in many social animals—wolves, for example—and it seems plausible that dogs exhibit a subset of wolf behavioral strategies, the ones that worked well under domestication. If some wolves are genetically inclined to try to become pack leaders, others are probably natural followers, and dogs likely have higher frequencies of such "sidekick" characteristics. We expect that differences between human ethnic groups are qualitatively similar to those between dog breeds—that the differences are evolutionarily shallow, mostly involving loss of function, exaggerations of already-existing adaptations, neoteny, and so on. Although such changes cannot generate truly complex adaptations, changes in all those hundreds or thousands of genetic switches and knobs can still cause the sorts of evolutionary changes we see in dogs and other domesticated species; and these differences—such as those between Great Danes and Chihuahuas, or between teosinte and modern maize—are not so small. In other words, very significant evolutionary changes in response to agriculture were still possible.

Not only are there strong reasons to believe that significant human evolution over the past 50,000 years is theoretically possible, and in fact likely, but it's completely obvious that it has taken place, since people look different. This is especially true of populations separated by great distances and geographical barriers. These differences are often so great that there is high contrast in appearance between populations: No Finn could be mistaken for a Zulu, no Zulu for a Finn. Differences in appearance have a genetic explanation, so we know that there has been substantial genetic change since modern humans spread out of Africa—change that has not taken the same course in every population.

It has been said that the differences between human populations are superficial, consisting of surface characteristics such as skin color and hair color rather than changes in liver function or brain development. In a letter to Vince Sarich and Frank Miele in eSkeptic, the e-mail newsletter of the Skeptics Society, Chuck Lemme said that "our insides do not vary like our outsides" and that differences in appearance are only skin-deep.[12] Lemme thinks that these superficial differences are probably driven by sexual selection, which would make them rather like fads.[13] Of course, since experts can easily determine race from skeletal features, it appears that those skin-deep differences go all the way to the bone. In fact, recent work has shown that there *are* population differences in genes affecting brain development, which we'll mention in Chapter 4.

It was natural for previous generations of physical anthropologists to concentrate on differences in easily observed characteristics, but that never implied that all differences would be easily observable. It was the scientists that were superficial, not the differences.

Some argue that differences between human populations are small and not very significant. As was eloquently pointed out by Richard Lewontin in 1972, most genetic differences are found within human populations rather than between different groups. Approximately 85 percent of human genetic variation is within-group rather than between groups, while 15 percent is between-group. Lewontin and others have argued that this means that the genetic differences between human populations must be smaller than differences within human population groups.[14] But genetic variation is distributed in a similar way in dogs: 70 percent of genetic variation is within-breed, while 30 percent is between-breed. Using the same reasoning that Lewontin applied in his argument about human populations, one would have to conclude that differences between individual Great Danes must be greater than the average difference between Great Danes and Chihuahuas. But this is a conclusion that we are unable to swallow.

It turns out that although the distribution of genetic variation is as Lewontin said, his interpretation was incorrect. Information about the distribution of genetic variation tells you essentially nothing about the size or significance of trait differences. The actual differences we observe in height, weight, strength, speed, skin color, and so on are real: It is not possible to argue them away. Genetic statistics do not tell you what sort of differences in size, strength, life span, or disposition you can expect to see between populations.

It turns out that the *correlations* between these genetic differences matter. If between-group genetic differences tend to push in a particular direction—tend to favor a certain trend—they can add up and have large effects. For example, there are undoubtedly a number of genes that affect growth in dogs, in the sense that some variants of those genes enhance growth and others inhibit it. Even if we find pro- and anti-growth gene variants in both Great Danes and Chihuahuas, the trend must be different. Growth-favoring variants must be more common in Great Danes. Even though a particular Great Dane may have a low-growth version of a particular gene, while a particular Chihuahua has the high-growth version, the sum of the effects of many genes will almost certainly favor greater growth in the Great Dane. We feel safe in saying this, since as far as we know, no adult Chihuahua has ever been as big as any adult Great Dane. In just the same way, on a given day it may rain more in Albuquerque, New Mexico, than in Hilo, Hawaii—but over the course of a year, Hilo is almost certain to be wetter. This has been the case for every year for which we have records.

More to the point, consider malaria resistance in northern Europeans and central Africans. Someone from Nigeria may have the sickle-cell mutation (a known defense against falciparum malaria), while hardly anyone from northern Europe does, but even the majority of Nigerians who don't carry sickle cell are far more resistant to malaria than any Swede. They have malaria-defense versions of many genes. That is the typical pattern you get from natural selection—correlated changes in a population, change in the same general direction, all a response to the same selection pressure.

For that matter, changes in a single gene can occasionally have a large effect: We know that terrifying genetic diseases can be caused by a change to a single gene, and we know that some of the key changes that occur in domestication are caused by mutations in a single gene.

For example, wild almonds contain amygdalin, a bitter chemical that turns into cyanide when the almonds are eaten. Eating a few wild almonds can be lethal. But in domesticated almond trees, mutations in a single gene block the synthesis of amygdalin, making the almonds edible.[15]

Such dramatic consequences of small genetic changes are possible because DNA is a bit like a recipe or a computer program: A change in a single letter can sometimes have a huge effect. In a striking example, the most common kind of dwarfism is caused by a change in a single nucleotide, rather like the meaning of an entire book changing because of one typographical error. In principle, differences in a single gene could cause significant trait differences between human populations.

The effect of genetic differences on body and mind must depend on the importance of the effects of the genes that vary between populations compared to those that vary within populations. Variants that have large effects will matter more than those that have small effects, right? Lewontin's argument assumed that the average impact of variants in those two classes was the same, which is incorrect. Since all humans have a fairly recent common ancestry (≈100,000 years), while humans outside of Africa have an even more recent common ancestry (≈50,000 years), observable differences between populations must have evolved rapidly, which can only have happened if the alleles (gene variants) underlying those differences had strong selective advantages. The alleles that are regional, those underlying the differences between populations, must also have had important effects on fitness. That's what population genetics implies, and genomic information now confirms it. Most or all of the alleles that are responsible for obvious differences in appearance between populations—such as the gene variants causing light skin color or blue eyes—have undergone strong selection. In these cases, a "big effect" on fitness means anything from a 2 or 3 percent increase on up. Judging from the rate at which new alleles have increased in frequency, this must be the case for genes that determine skin color (SLC24A5), eye color (HERC2), lactose tolerance (LCT), and dry earwax (ABCC11), of all things.[16]

In many cases common ancestry is even more recent—for example, Amerindians and northern Asians appear to have diverged only 15,000 years ago or thereabouts. In these populations, selection has had even less time to operate, and observed differences must have had even bigger impacts on fitness.

Thus, we believe that the obvious differences between racial groups are linked to gene variants that have *recently* increased in frequency and had major fitness effects. Blue eyes, found only in Europeans and their near neighbors, are the result of a new version of the DNA that controls the expression of OCA2 that has undergone strong selection, at least in Europe. Dry earwax is common in China and Korea, rare in Europe, unknown in Africa: The gene variant underlying dry earwax is the product of strong recent selection. We can confidently predict that many (perhaps most) as yet unexplained racial differences are also the product of recent selection. For example, we argue that the epicanthic eyelid fold found in the populations of northern Asia is most likely the product of strong and recent selection.

All this means that just as humans 40,000 years ago were significantly different from their ancestors 100,000 years ago (much more inventive, in particular), humans today are different in many ways from our ancestors of 40,000 BC, and, considering the accelerated rate of change, different from our ancestors of early historical times as well. We can empathize with the heroes of the *Iliad* (well, Odysseus at any rate)—but we're not the same.

Before the development of modern molecular biology, there were severe limitations on our ability to study human evolution. Then, all we had to work with were the principles of genetics, easily observable differences between peoples such as skin color, and detailed knowledge of a limited number of genes, mostly blood proteins and those causing genetic diseases such as sickle-cell anemia.

But even then, we knew from our experience with animal and plant breeding, along with observation of many examples of rapid evolution in nature, that there could be significant evolutionary change in 10,000 years or less. It was also clear that modest genetic differences between groups could cause big trait differences. Indeed, entirely divergent life strategies can be caused by differences in a single gene, as we see in fire ants, where ants with one version of a pheromone receptor live in independent colonies, each having a single queen, while those with the other version live in a sprawling metacolony with many queens.[17] Well before the revolution in genomics, it was clear enough that there could be significant differences between human populations in almost any trait, despite recent common ancestry. It was clear that this was entirely compatible with what we knew of genetics, and it was also clear that at least some such differences existed in skin color, size, morphology, and metabolism.

But as the molecular revolution has unfolded over the past few years, we have learned a great deal more. Recent studies have shown that many genes are currently being replaced by new variants, most strongly in Eurasians—and that those genes favored by recent selection are for the most part different in different populations. The obvious between-population differences that we knew of a few years ago were only the tip of the iceberg.

## LINKAGE

Most of the recent studies have used data from the HapMap, a database of common patterns of human genetic variation produced by an international group of researchers. Four populations were selected for the HapMap—ninety Nigerians, ninety Americans of European ancestry, forty-five individuals from Tokyo, and forty-five from Beijing. For some purposes we will group the Japanese and Chinese together as an "East Asian" sample.

The human genome has about 3 billion *bases* (the four molecular building blocks that make up DNA) organized into forty-six separate bundles of DNA called *chromosomes*. For the most part, DNA sequences are the same in all humans, but every few hundred bases, a variable site crops up. These are the only sites in which the bases of DNA are likely to vary from one individual to another.

A particular pattern of variation at these sites is called a *haplotype*. Imagine three successive variable sites—the first can be G or C (representing guanine or cytosine), the second can be A (adenine) or G, and the third can be T (thymine) or C. A particular individual might have C in the first site, A in the second site, and T in the third site—his haplotype would be CAT—while another person has the haplotype cytosine-guanine-thymine, or CGT. A haplotype is like a hand of poker, while the bases in the variable sites are like individual cards.

And just like cards, haplotypes are shuffled. In each generation, a new chromosome is assembled from the inherited parental chromosomes in much the same way that one can cut two decks of cards and assemble a new deck, a process we call *recombination*. There can be multiple cuts: in humans, an average of one to three per chromosome.

This means that haplotypes are partially broken down every generation: The complete pattern that existed over the whole of the parent's chromosome will no longer be intact after recombination. However, smaller parts of that pattern are likely to remain unchanged, since a chromosome is millions of bases long and the few breaks that occur are likely to be far away.

Over many generations, any haplotype will eventually be completely reshuffled. But if a favorable mutation occurs on a chromosome, people with that mutation will have more children survive than average, so over time, more and more people will bear that mutation. If the advantage is large enough, the mutation can rapidly become common, before recombination completely reshuffles its original haplotype, rapidly enough that people bearing that mutation will also carry the original local haplotype that surrounded it when it first came into existence. The longer the shared haplotype, the younger the mutation. It's as if part of your last hand of cards showed up again in the new deal: You would guess that there hadn't been much shuffling, and you'd be right.

The HapMap studies looked for long haplotypes (long unshuffled regions) that existed in a number of individuals in the dataset. Any such shared pattern would be a sign of recent strong selection—quite recent, since recombination eventually breaks down all such patterns.

One well-known example is the gene that makes lactase, the enzyme that digests milk sugar. In most humans, and in mammals generally, lactase production stops around the age of weaning, but in many Europeans and some other peoples, production continues throughout life. This adaptation lets adults drink milk. Lactose-tolerant Europeans carry a particular mutation that is only a few thousand years old, and so those Europeans also carry much of the original haplotype. In fact, the shared haplotype around that mutation is over 1 million bases long.

Recent studies have found hundreds of cases of long haplotypes indicating recent selection: Some have almost reached 100 percent frequency, more have intermediate frequencies, and most are regional. Many are *very* recent: The rate of origination peaks at about 5,500 years ago in the European and Chinese samples, and at about 8,500 years ago in the African sample. Again and again over the past few thousand years, a favorable mutation has occurred in some individual and spread widely, until a significant fraction of the human race now bears that mutated allele. Sometimes almost everyone in a large geographic region, such as Europe or East Asia, shares a trait that goes back to one such allele. The mutation can affect many different things—skin color, metabolism, defense against infectious disease, central nervous system features, and any number of other traits and functions.

Since we have sequenced the chimpanzee genome, we know the size of the genetic difference between chimps and humans. Since we also have decent estimates of the length of time since the two species split, we know the long-term rate of genetic change. The rate of change over the past few thousand years is far greater than this long-term rate over the past few million years, on the order of 100 times greater. If humans had always been evolving this rapidly, the genetic difference between us and chimpanzees would be far larger than it actually is.[18]

In addition, we see far more recent alleles at moderate frequencies (20 percent to 70 percent) than we do with frequencies close to 100 percent. Since a new favored allele spends a long time at low frequencies (starting with a single copy), a short time at moderate frequencies, and then a long time closing in on 100 percent, the only explanation is that this rush of selection began quite recently, so that few selected genes are in that final phase of increase.

The ultimate cause of this accelerated evolution was the set of genetic changes that led to *an increased ability to innovate.* Sophisticated language abilities may well have been the key. We would say that the new alleles (the product of mutation and/or genetic introgression) that led to this increase in creativity were *gateway mutations* because innovations they made possible led to further evolutionary change, just as the development of the first simple insect wings eventually led to bees, butterflies, and an inordinate number of beetles.

Every major innovation led to new selective pressures, which led to more evolutionary change, and the most spectacular of those innovations was the development of agriculture.

# 2 THE NEANDERTHAL WITHIN

In their expansion out of Africa, modern humans encountered and eventually displaced archaic humans such as Neanderthals: You can't make an omelet without breaking eggs. Moderns showed up in Europe about 40,000 years ago, arriving first in areas to the east and north of Neanderthal territory, the mammoth steppe that Neanderthals had failed to settle permanently. A superior toolkit—in particular, needles for sewing clothes—may have made this possible.

Later, modern humans moved south and west, displacing the Neanderthals. This is more or less what one would expect to happen, since the two sister species were competing for the same kinds of resources—ecological theory says that one will win out over the other. It took just 10,000 years for modern humans to completely replace Neanderthals, with the last Neanderthals probably living in what is now southern Spain.

Judging by outcomes, modern humans were competitively superior to Neanderthals, but we don't know what their key advantage was, any more than we know what drove the expansion of modern humans out of Africa. Several explanations have been suggested, and some or all of them may be correct.

One idea is that modern humans had projectile weapons, in contrast to the thrusting spears used previously. If lightly built modern humans could hunt just as well as Neanderthals while requiring fewer calories, strongly built Neanderthals would have become obsolete. Even if Neanderthals had managed to copy that technology, they would have expended more energy in hunts because of their heavier bodies. Finds of small stone points in the Aurignacian (a culture that existed in Europe between 32,000 and 26,000 BC) suggest that something like this scenario may have occurred, but the earliest known spearthrowers, or atlatls, were made considerably later. Another idea is that modern humans were smarter—which might have been the case, but it is hard to prove.

Probably the most popular and attractive hypothesis is that modern humans had developed advanced language capabilities and therefore were able to *talk* the Neanderthals to death. This idea has a lot going for it. It's easy to imagine ways in which superior language abilities could have conferred advantages, particularly at the level of the band or tribe. For example, hunter-gatherers today are well known for having a deep knowledge of the local landscape and of the appearance and properties of many local plants and animals. This includes knowledge of rare but important events that happened more than a human lifetime ago, which may have been particularly important in the unstable climate of the Ice Age. It is hard to see how that kind of information transmission across generations would be possible in the absence of sophisticated language. Without it, there may have been distinct limits on cultural complexity, which, among other things, would have meant limits on the sophistication of tools and weapons.

Beginning in Africa, and continuing in the European archaeological record, we see signs of long-distance trade and exchange among modern and almost-modern humans in the form of stone tools made out of materials that originated far away. The Neanderthals never did this: To the extent that such trade was advantageous, it would have favored moderns over Neanderthals, and it is easy to imagine how enhanced language abilities would have favored trade. Those trade contacts (and the underlying language ability) might have allowed the formation of large-scale alliances (alliances of tribes), and societies with trade and alliances would have prevailed over opponents that couldn't organize in the same way.

Whatever the driving forces, this population replacement was slow, at least when compared to the time scale of recorded history, and was most likely undramatic. The distance from Moscow to Madrid is a little over 2,000 miles; that's not a lot of ground to cover in 10,000 years. The actual advance of modern humans in Europe may have taken the form of occasional skirmishes in which moderns won more often than not. Perhaps modern humans were better hunters and made big game scarce, so that neighboring Neanderthal bands suffered. Perhaps moderns, with their less bulky bodies and more varied diet (including fish), were better at surviving hard times. Quite possibly, the actual advance was made up of a mix of all these patterns.

Of course, there are other possibilities. Biological advantages take many forms, and they don't have to be admirable—in fact, they can be downright embarrassing, disgusting, or, worst of all, boring. One realistic and embarrassing possibility is that modern humans expanding out of Africa carried with them some disease or parasite that was fairly harmless to them but deadly to Neanderthals and the hominid populations of East Asia—the "cootie" theory. There is no direct evidence for this, but then it's hard to see how there would be: Germs seldom leave fossils. We know of natural examples of this mechanism, however. White-tailed deer carry a brain worm that is fairly harmless to them but fatal to moose.[1] So white-tailed deer are pretty good at displacing moose populations and have been doing so since their traditional enemies, such as wolves, have mostly disappeared. Another example: Since people imported American gray squirrels to England, the native red squirrel has declined dramatically. The gray squirrels carry a virus that they survive but that devastates the native red squirrels.[2]

We have heard charmingly goofy criticisms of the idea that Neanderthals were competitively inferior to modern humans. It has been suggested that such a position is racist. Somehow, saying that a population that split off from modern humans half a million years ago (one generally considered a separate species), had some kind of biological disadvantage is beyond the pale, even though we're here and Neanderthals are not. For that matter, we've seen people argue that the idea that some genes were picked up from archaic humans is racist, while others have argued that the idea that humans *didn't* pick up Neanderthal genes is racist.

Although the archaeological evidence suggests that moderns and Neanderthals did not coexist for very long in any one place during this replacement, there is reason to believe there was some contact between the two different populations. In several places, most clearly in central and southwestern France and part of northern Spain, we find a tool tradition that lasted from about 35,000 to 28,000 years ago (the Châtelperronian) that appears to combine some of the techniques of the Neanderthals (the Mousterian industry) with those of modern humans (the Aurignacian). The name Châtelperronian comes from the site of the Grotte des Fées (Fairy Grotto) near the French town of Châtelperron. Châtelperronian deposits contain flint cores characteristic of the Neanderthals' Mousterian technology mixed with more modern tools. One characteristic tool was a flint knife with a single cutting edge, in contrast with the double-edged knives we see in the Aurignacian industry. Most important, there are several skeletons clearly associated with the Châtelperronian industry, and all are Neanderthal. This strongly suggests that there were interactions between the populations, enough that the Neanderthals learned some useful techniques from modern humans. If this is the case, it tells us something about Neanderthal cognitive capabilities—mainly that they can't have been all *that* far behind modern humans. At minimum, they were much better at learning new things than chimpanzees.

There may have been important consequences from such interactions; familiarity may breed contempt, but lack of familiarity breeds nothing at all.

## THE "BIG BANG"

"The Upper Paleolithic," according to Stanford anthropologist Richard Klein, "signals the most fundamental change in human behavior that the archaeological record may ever reveal, barring only the primeval development of those human traits that made archaeology possible."[3]

He's not kidding. The archaeological record of the Upper Paleolithic, or last phase of the Old Stone Age—the product of the modern humans who displaced the Neanderthals in Europe 30,000 to 40,000 years ago—is qualitatively different from anything that came before. With the advent of modern humans in Europe, innovation was bustin' out all over.

Many of the new features that marked this "great leap forward" were impressive—cave paintings, sculpture, jewelry, dramatically improved tools and weapons. Some of them brought significant changes in the practical matters of daily life, but most important, from our point of view, is that they show an extraordinary increase in the human capacity to create and invent.

What's more, the innovations show that profound social and cultural changes were taking place. We developed new social arrangements as well as new tools: The spearpoints and scrapers of this period often used materials from hundreds of miles away, which must have been acquired through some form of trade or exchange. Before, tools were almost entirely made from local materials. We also see the beginnings of cultural variation: Tools and weapons started showing regional styles.

At this point, people—some of them anyhow—were acting wildly different from their forebears of even 20,000 years earlier. The spark of innovation was taking them in all kinds of new directions. We're not saying that every Tom, Dick, and Harry was an inventor, but at least some people were coming up with new ideas—and doing so perhaps 100 times more often than in earlier times. The natural question is, "Why?" It doesn't really look as if being a modern human, in the sense of having ancestors who were anatomically modern and who had originated in Africa, was enough, by itself, to trigger this change. We don't see this storm of innovation in Australia. Obviously, something important, some genetic change, occurred in Africa that allowed moderns to expand out of Africa and supplant archaic sapiens. Equally obviously, judging from the patchy transition to full behavioral modernity, there was more to the story than that. So probably being an "anatomically modern" human was a necessary but not sufficient condition for full behavioral modernity.

More generally, behavior has a physical substrate: Biology keeps culture on a leash, which is why you can't teach a dog to play poker, never mind all those lying paintings. We have every reason to think that back in the Eemian period (the interglacial period of about 125,000 years ago), the leash was too short for agriculture. Humans did not develop agriculture anywhere on earth during the Eemian, but they did so at least seven times independently in the Holocene, the most recent interglacial period, which began 10,000 years ago. Not only that, in the Eemian the leash was too short to allow for the expansion of anatomically modern humans out of Africa into cooler climates. In that period, biology somehow kept people from making atlatls or bows, and from sewing clothes or painting, all of which are routinely performed and highly valued by contemporary hunter-gatherers. People were different back then—significantly different, biologically different.

Genetic changes allowed important human developments in 40,000 BC that hadn't been possible in 100,000 BC. Moreover, other genetic changes may have been necessary precursors to later cultural changes. Here we shall argue that the dramatic cultural changes that took place in the Upper Paleolithic, which have been referred to as the "human revolution," the "cultural explosion," or (our favorite), the "big bang," occurred largely because of underlying biological change.

We are not the first to suggest this. Richard Klein has said that some mutation must have been responsible for this dramatic increase in cultural complexity.[4] We wholly agree with the spirit of his suggestion, but we believe that such dramatic change probably involved a number of genes, and thus some mechanism that could cause unusually rapid genetic change. As it turns out, we know of such a mechanism, and the necessary circumstances for that mechanism turn out to have arrived just in time for the human revolution.

## NEW AND IMPROVED

So what exactly were the innovations of the Upper Paleolithic that have drawn attention to this period as a time of revolutionary change? For one thing, we see new tools, made from new materials—tools made using careful, multistep preparation. Modern humans still used stone (although their methods of preparation had grown more elaborate and efficient), but they often used bone and ivory as well, in sharp contrast with Neanderthals. They also used particular types of high-quality stone from distant sources, sometimes from as far as hundreds of miles away, a pattern that suggests trade. New types of light, high-velocity weapons appeared, such as javelins, atlatl darts, and eventually the bow and arrow. These weapons, which could be used at a distance, must have made bringing down big game far safer than it had been among hunters using thrusting spears. The skeletons of modern humans in this period, unlike those of

Neanderthals, are not thoroughly beat up. The weapons were no doubt used for warfare and defense, but they were primarily for hunting, and their benefits included broadening the modern humans' diet.

Moderns hunted small game and fish, in addition to the large game favored by their predecessors. This more varied diet (and perhaps safer hunting methods) led to higher population density—archaeological sites for modern humans during this period become several times more common than they had been among the Neanderthals. The moderns were able to catch fish using newly devised tools such as fishhooks, nets, and multi-barbed harpoon points. Those nets are a manifestation of another technological innovation, the use of plant fibers to make baskets, textiles, and rope as well as nets and snares.

Moderns developed new methods of preserving food, such as using drying racks and pits dug in the permafrost, which acted as natural refrigerators. They employed fire more efficiently than their ancestors had, developed hearths that had draft channels for better air flow, and began to use warming stones for cooking. Fire was used in other specialized ways as well—in lamps, for example, and to make pottery figures.

Burial—deliberate burial with clear-cut evidence of ritual—also becomes much more common in the Upper Paleolithic. The remains are often accompanied by grave goods such as tools, shells, personal items of jewelry, and red ochre. In some cases, production of those grave goods took tremendous effort. In Sungir, near Moscow, individuals were buried in clothes decorated with thousands of ivory beads whose manufacture required several man-years of effort. These findings suggests a hierarchically differentiated society, with chiefs as well as Indians.

These elaborate burials are in sharp contrast to Neanderthal burials, which show no sign of ceremony. We don't find weapons or decorative objects associated with those graves. It may be that for the Neanderthals, burial was more a way of disposing of unpleasant remains than a ritual occasion, something like flushing a goldfish down the toilet.

Modern humans began to build much more substantial protective structures. At Dolni Vestonice, located in what is now the Czech Republic, archaeologists have found the remains of five structures marked by mammoth bones, blocks of limestone, and postholes, the largest covering more than 1,000 square feet. In Russia and Ukraine, where natural shelters such as limestone caves were scarce, we see dwellings that use many mammoth bones. Building them must have involved serious effort: One such house contained some 23 tons of the bones of these large mammals.

The most striking change of the Upper Paleolithic, to modern eyes, is the birth of art. The most spectacular examples are the cave paintings, found primarily in France and Spain. Typical subjects are large animals such as bison, deer, and aurochs, but sometimes predators such as lions, bears, and hyenas are depicted. Made with carbon black or ochre, these paintings usually depict animals naturalistically. Humans, which show up rarely, often look quite strange.

The first real sculptures also appeared during this time. The most famous, the Venus figurines, such as the famous Venus of Willendorf (see page 38), may have been portable pornography. At Dolni Vestonice, researchers found ceramic figures made about 29,000 years ago, long before the invention of pottery in other parts of the world.

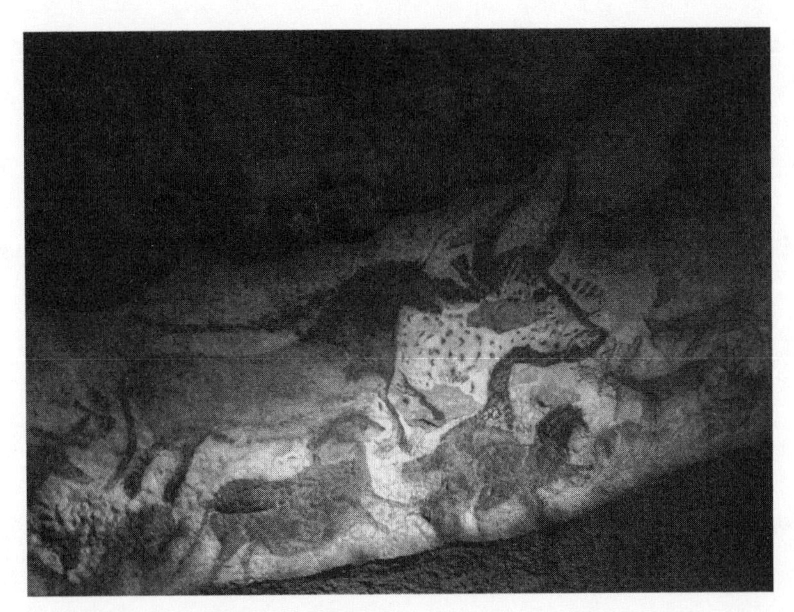

Lascaux cave painting, ≈14,000 BC

The art of the Upper Paleolithic was qualitatively different from the first symbolic objects seen in Africa before the expansion of modern humans: Compare the incised piece of ochre from Blombos Cave in South Africa dated about 75,000 BC[5]—representative of the most sophisticated symbolic objects discovered in pre-expansion Africa—with the lion-headed sculpture carved from mammoth ivory, found in Germany and dated about 30,000 BC (see pages 38 and 39).

## FUSION

The tremendous changes in tools, in weaponry and hunting methods, and in art, along with the social and cultural changes they imply, could not have simply come out of the blue. The Upper Paleolithic advances point to some underlying mechanism that generated rapid genetic changes that conferred new capabilities. That mechanism, we believe, was introgression—that is, the transfer of alleles from another species, in this case Neanderthals. There is no faster way of acquiring new and useful genes.

Before we go further, we must acknowledge that this idea has not been much considered by paleontologists and anthropologists, mainly because they are not familiar with the arguments derived from population genetics that show that such introgression is highly likely. In addition, members of the general public who hear it for the first time may well be put off by the idea, since Neanderthals are usually considered backward, even apelike.

Many object to the notion of humans and Neanderthals mating and having offspring. Their first impulse is to suggest that anatomically modern humans and Neanderthals must have been too different, so that matings would not have produced fertile offspring. They say that humans would never have done such a disgusting thing. And they say that even if it happened, it was almost certainly rare, and thus biologically insignificant. None of these claims are correct: We will address them all.

The issue of whether or not there was mating between modern humans and Neanderthals is central to the debate that has raged for several decades about multiregional evolution versus a single African origin of our species. The strong multiregional position held that Neanderthals were directly ancestral to humans,[6] while the strong single-Africa-origin model held that modern humans simply replaced the Neanderthals.[7] It quickly became apparent in the face of genetic data that a dramatic out-of-Africa dispersal of modern humans did occur, but the extent of genetic exchange between the old and new humans was not resolved. Much debate occurred about whether there were anatomical continuities between Neanderthals and contemporary Europeans, the underlying assumption being that some sort of anatomical blending would have occurred. Our perspective on the issue, elaborated below, is quite different.

### Interfertility

The first point made by critics is that modern humans and Neanderthals could not have been interfertile. However, we believe that they almost certainly were, since the two species had separated fairly recently, roughly half a million years earlier. No primates are known to have established reproductive isolation in so short a time.[8] Bonobos, for example, branched off from common chimpanzees some 800,000 years ago, but the two species can have fertile offspring.[9] Most mammalian sister species retain the ability to interbreed for far longer periods.[10] Sometimes zookeepers are surprised by this, as when a dolphin and a false killer whale produce viable offspring.[11] There are rumors about successful matings between primate lineages that separated as long as 5 million or 6 million years ago, but those are currently unsubstantiated. Nevertheless, there is no reason to think that during the Upper Paleolithic Neanderthals and anatomically modern humans could not have mated and had children who lived to also reproduce.

### Bestiality?

As for the idea that people just wouldn't have wanted to mate with creatures that were so different, we can only say that humans are known to have had sexual congress with vacuum cleaners, inflatable dolls, horses, and the Indus river dolphin. Any port in a storm, as it were. Jared Diamond recounted how a physician friend, treating a pneumonia patient with a limited

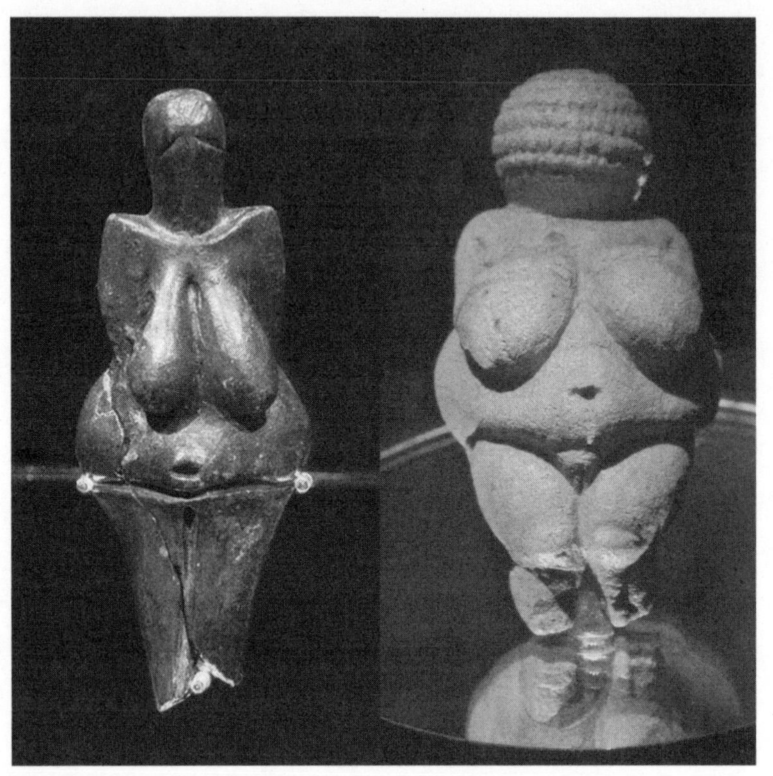

Venus of Dolni Vestonice, oldest known ceramic, ≈27,000 BC

Venus of Willendorf, ≈23,000 BC

Blombos ochre, one of the oldest known symbolic objects, ≈70,000 BC

Lion Man of Hohlenstein, oldest known animal sculpture, ≈30,000 BC

command of English, had the patient's wife ask him if he'd had any sexual experiences that could have caused the infection. After the man recovered consciousness (his wife had knocked him cold as he began to answer), he admitted to repeated intercourse with sheep on the family farm.

The key point, which we will show in more detail later on, is that even rare interbreeding can be very important. If someone wanted to show that interbreeding between Neanderthals and modern humans was biologically insignificant, he would have to show that it never happened—and that is most unlikely, considering the human track record. If it happened at all, then introgression could have had a huge impact on human development.

### 2s

A number of researchers have suggested that matings between Neanderthals and modern humans were rare and therefore biologically unimportant.[12] But this objection is definitely incorrect: It is based on a misunderstanding of the genetics of natural selection. Some anthropologists who study anatomical details of Neanderthals and modern humans see evidence of Neanderthal features in some of the earliest modern humans in Europe,[13] but others dispute the matter.

Imagine that humans occasionally mated with Neanderthals, and that at least some of their offspring were incorporated into human populations. That process would have introduced new gene variants, new alleles, into the human population. Many, probably most, of those alleles would have done almost exactly the same thing as their equivalents in modern out-of-Africa humans; they would have been neither better nor worse than those equivalents—in other words, they would have been selectively neutral. Those neutral alleles from Neanderthals would have been rare, and they would probably have disappeared, the typical fate of rare neutral alleles.

The reason is simply chance. When the bearer of a rare neutral allele has a child, that child has a 50 percent chance of carrying that allele. With two children (the average number in a stable population), there's a 25 percent chance that neither child will have a copy, and in that case, the imported allele disappears right then and there. More generally, the number of copies of a neutral allele fluctuates randomly with time, and any time the number hits zero, the story ends. If the original number of copies is low, this is fairly likely. Even if, by sheer luck, one or two neutral Neanderthal alleles had eventually become common in modern humans, there would have been no real consequences, since neutral alleles are boring by definition. Neanderthal alleles with negative consequences (in humans) would have disappeared even more rapidly. But some gene variants provide biological advantages and are adaptive. For those advantageous alleles, the story is entirely different.

The key property of an advantageous allele is that its frequency tends to increase with time, usually because it aids the bearer in some way. In a stable population, this means that the number of copies in the next generation is (on average) larger than the number in the current generation. If the average number of copies in the next generation were one and a quarter times larger than in the first, we would say that the allele had a selective advantage of 25 percent. As favorable alleles go, 25 percent is a very large advantage, although not unprecedented.

A single copy of an advantageous allele can still disappear, and probably will. With a 10 percent fitness advantage, a carrier in an otherwise stable population would average 2.2 offspring instead of 2, and there would still be a 23.75 percent chance of that allele disappearing in the first generation. But there is a way in which copies of this allele can survive: If luck holds out long enough, they will become more common—eventually, so common as to be effectively immune to chance. From that point on they steadily increase in numbers.

J. B. S. Haldane, the great British geneticist (1892–1964), found a systematic way of adding up all these probabilities, and his method yields a surprisingly simple answer. If the allele confers an advantage $s$, its chance of going all the way is $2s$. In a stable population, a single copy of an allele with a 10 percent fitness advantage has a 20 percent chance of eventually becoming universal.

The fate of one copy of a favorable allele is very much like that of a gambler who starts out with one chip and a roulette system—a way of beating the odds—that really works. If he can pick the correct color (red or black) 55 percent of the time and bet one chip at a time, he'll usually go broke—but there's an 18 percent chance that he'll break the bank at Monte Carlo. And that's starting with one chip. With twenty chips, our friend (and who wouldn't want to have a friend like this?) would have a 98 percent shot at victory.

What this means is that one copy of an advantageous allele is much more likely to reach high frequencies than a single copy of a neutral allele—so much so that even a few dozen half-Neanderthal babies over thousands of years would have allowed modern humans to acquire most of the Neanderthals' genetic strengths.

Let's sketch an example. A neutral allele's chance of drifting to 100 percent (a state called "fixation") is the inverse of the number of gene copies in the population—one divided by twice the number of breeding individuals in the population, since each individual carries two copies of that gene. In other words, a neutral copy has exactly the same chance of reaching high frequency as every other neutral copy of that gene. For a population of any size, that chance is very small—for example, a chance of 1 in 20,000 for a human population with an effective size of 10,000. Such drift is also a very slow process, usually taking tens of thousands of generations.

Now consider an advantageous allele—a single copy of a new and improved version of a gene involving the immune system, one that made the bearer immune to some common and dangerous disease that normally killed off 10 percent of the population in childhood. That new allele would have a selective advantage of 10 percent. It might vanish; in fact it probably would, if, for example, the bearer managed to be stepped on by a mammoth or if none of his or her kids happened to carry that gene. But barring such accidents, the number of copies of that gene would tend to increase. Once the number of copies reached 50 or 100, the gene would be very unlikely to disappear by chance. From that point on there would be a fairly steady increase. It turns out that a single copy of that gene would have a 20 percent chance of making it big—going from one individual to, eventually, a significant fraction of the human race over the course of a few thousand years—assuming that the advantage persisted. That is, it would be 4,000 times more likely than a single neutral allele to reach fixation, and the process would be much faster.

If this advantageous allele was introduced by hybridizing with another species, rather than as a new mutation, it would likely be introduced repeatedly over a relatively short period of time, since there would probably be a number of such matings. If ten copies were introduced, the odds would be high that at least one of those copies would become a big success.

This reasoning goes against our intuition. Generally, we think that ancestry is something like mixing colors of paint: If you pour in equal amounts of blue and yellow, you'll get green—and the paint will remain green. If a population were 90 percent Norwegian and 10 percent Nigerian, intuition says that nine-to-one mix will remain the case indefinitely. But intuition is wrong: If you placed that mixed population in Africa, certain alleles that were common in Nigerians—alleles that protected against malaria, or that made skin dark and resistant to skin cancer—would become more and more common over many generations. Eventually almost everyone in that population would carry the Nigerian version of those genes.

In just this way, a tiny bit of Neanderthal ancestry thrown into the mix tens of thousands of years ago could have resulted in many people today, possibly even all modern humans, carrying the advantageous Neanderthal version of some genes.

## HOW DID IT HAPPEN?

If there really was interbreeding between Neanderthals and anatomically modern humans, how and where might it have happened?

There certainly might have been some gene flow among hominid species in earlier times. After all, *Homo heidelbergensis* (the common ancestor of Neanderthals and anatomically modern humans) somehow managed to settle both Europe and Africa about half a million years ago, so communication must have been possible, at least occasionally. It may have been impossible most of the time because of the Sahara Desert, which is a potent barrier today, just as it was during the ice ages. The Sinai Peninsula, also often a desert in its history, may also have been an important barrier, since it was the only land connection between Africa and Eurasia. More than that, Neanderthal alleles that were advantageous outside of Africa may not have been so in Africa and thus might not have spread to anatomically modern humans there.

We have reason to think that the modern humans who expanded out of Africa some 50,000 years ago had changed in important ways—had, for example, probably acquired sophisticated language abilities. A Neanderthal allele that had not been particularly useful in the genetic context of near-modern humans 100,000 years ago might have been useful to the more advanced people who were expanding out of Africa.

Logically, if admixture occurred at all, it had to happen somewhere in Neanderthal-occupied territory, which means Europe and western Asia. As modern humans expanded their territory, they must have encountered Neanderthal bands again and again. The two kinds of humans coexisted for a few thousand years before the Neanderthals disappeared, at least in some regions. This looks to be the case for the Châtelperronian culture of France and northern Spain, and there are traces of a similar culture in Italy. If there was trade, or if there was enough contact to transmit toolmaking techniques, there was sexual contact as well—depend on it. If in the future we look at very large genetic datasets from huge numbers of individuals, we might find a few traces of neutral Neanderthal genes.[14]

If we found a few individuals with Neanderthalish mitochondrial DNA (mtDNA) or Y chromosomes, we might be able to determine whether matings occurred mostly between

Neanderthal males and modern females or modern males and Neanderthal females. We do this type of analysis routinely today and have found, for example, that the maternal ancestry for most Mexicans is Amerindian, whereas most of their paternal ancestry is Spanish (which simply means that male Spanish explorers sometimes mated with Amerindian women). At this point, though we have found no Neanderthal Y chromosomes or mtDNA in modern humans, we cannot rule out significant introgression, because Neanderthal mtDNA and Y chromosomes may well have been neutral or deleterious in modern humans. In either case they would be unlikely to have persisted until today, particularly if the amount of gene flow was small. This does not mean we did not inherit beneficial gene sequences (see section on "Genetic Evidence" later in this chapter).

## BUT I DON'T WANT TO BE PART NEANDERTHAL!

There is often a visceral reaction to the idea that we carry some Neanderthal genes. Probably this is due to the general impression that Neanderthals were backward and apelike. Neanderthals weren't really apelike, although they were behind the times—but since it looks, in any case, as if we've absorbed only their best (most useful) traits, we can be happy about our Neanderthal ancestry, proud even. At any rate, it could be worse: We could have picked up genes from a virus. In fact, it *is* worse: We have.

Most viruses (which are basically just bags full of DNA or RNA) slip into cells and then take over, making copies of themselves and usually killing the host cells in the process. But some RNA viruses (retroviruses, like HIV) copy their RNA into DNA and then, sometimes, integrate that DNA into the host

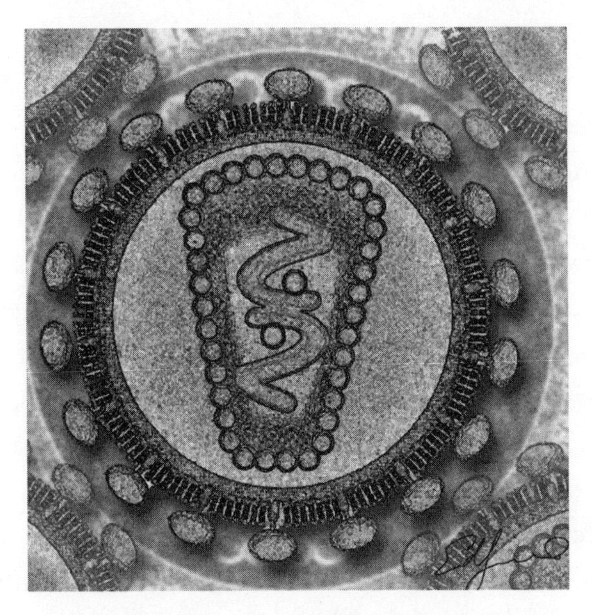

Stylized rendering of the AIDS retrovirus

cell's genome. If the retrovirus happens to occupy a reproductive cell, one that makes sperm or eggs, the retroviral genes can actually become part of the next generation's genome. This has happened in the past: Humans have many genetic remnants of retroviruses that at one time inserted copies of themselves into the human genome. Most do not seem to have any real function, but a few do. For example, both humans and apes have syncytin, derived from a retroviral envelope protein that our ancestors picked up roughly 30 million years ago. It plays a role in the development of the placenta—in particular, the process that leads to the development of a fused cell layer. Anyone who's overly worried about possible Neanderthal ancestry should remember that we're certainly descended from viruses. As usual, the facts don't care about our feelings.

## EXAMPLES OF INTROGRESSION

Introgression as an important evolutionary force is more than just a theory: Geneticists know of many cases in which it has definitely occurred. Most of the examples that are well understood involve domesticated animals and plants, mainly because there are practical economic reasons for undertaking close genetic studies of domesticated species.

Introgression is hardly rare; in fact, it is ubiquitous among domesticated plants. For example, the wheat that produces our daily bread is derived from three different wild grasses. There's evidence of introgression in alfalfa, barley, chili peppers, lettuce, maize (corn), potatoes, rice, rye, sorghum, and soybeans—and that's just a partial list. But since plants are better than animals at tolerating complex genetic events such as changes in chromosome number, introgression in domesticated animals may be a better analogy.

Cows were domesticated at least twice: in the Middle East (humpless taurine cattle) and in India (humped, droopy-eared zebu cattle), and possibly a third time in North Africa. The wild ancestors of taurine and zebu cattle were separated for several hundred thousand years, yet those breeds are interfertile. Zebu genes have been spreading among taurine cattle in Africa and western Asia for the past 4,000 years. It appears that some zebu genes increase tolerance of aridity and heat as well as resistance to rinderpest, a virulent bovine disease. This is very similar to the pattern of introgression we believe must have occurred among moderns and archaics such as Neanderthals.

Evidence of adaptive introgression in wild populations was once rare, but it has become easier to discover and document it

Zebu cow

Texas longhorn

in recent years thanks to improved DNA-sequencing techniques. We now have genetic evidence for adaptive introgression in wild organisms such as damselflies, mosquitoes, lake trout, and European hares. One of the most interesting cases of the phenomenon in a (partly) wild population is the recent evolution of honey bees, which is particularly interesting because of some close parallels with human evolution.

Honey bees originated in eastern tropical Africa several million years ago and later expanded into Eurasia in two different migrations. One of these led to Western European honey bees and the other to Asian honey bees. Bees living in temperate climates faced fundamentally new problems: more than anything else, cold winters. To a large extent, their adaptation to those new climates was mediated by changes in social behavior.

They needed to choose nest sites that would protect them from the weather, store much more honey, and form a winter cluster—that is, a tightly packed clump of bees that conserves heat. In a 2008 study, Amos Zayed and Charles Whitfield concluded that approximately 10 percent of all protein-coding genes in bees underwent positive selection in that process of adaptation.[15] The history of honeybees parallels the history of humans in an interesting way—both involved expansion into a new environment with a drastically different climate followed by strong selection and adaptation.

The parallels don't stop there. Soon after the initial colonization of the Americas, Europeans introduced honey bees, where they did well, swarming many times a year and outrunning colonists. However, they did not do as well in the neotropics, the part of the New World most unlike Europe. Warwick Kerr, a twentieth-century Brazilian geneticist and bee breeder, attempted to develop a strain of bees that would be more productive in the tropics. Using hybridization techniques, he bred Western European bees with African bees. In 1956, twenty-six of his Tanzanian queens escaped and began colonies, and their hybrid descendants have since spread over much of North and South America. These Africanized bees produce more honey than European bees in warm climates, but they are very aggressive and often attack people and animals that come too close to their hives. Once aroused, they may chase an enemy a mile or more. This high level of aggression is adaptive in Africa, where bees have not been domesticated. There, bee colonies are attacked by honey badgers and other predators, and humans raid beehives rather than keeping bees.

Almost all Africanized bee colonies have African mtDNA, but a significant fraction of their nuclear genome is European. That fraction is significantly higher in coding regions of the genome than in noncoding DNA, which indicates that Africanized bees have succeeded in picking up adaptive alleles from European bees while retaining those African alleles that are adaptive in this situation (that is, most of them). The noncoding regions, presumably neutral or close to neutral, are incorporated at the rate $1/(2N)$, while favorable coding genes are incorporated at the $2s$ rate, as discussed above. It has been suggested that there may be genetic incompatibilities between European mtDNA and the African nuclear genome, which would explain why we find so very few Africanized bees with European mtDNA. The Zayed-Whitfield study provided evidence that "invasive populations can exploit hybridization in an adaptive fashion," which is only reasonable. Just as the Africanized bees have incorporated advantageous genes from the local indigenous bee populations, modern humans, we believe, incorporated advantageous genes from their archaic human precursors, especially from the Neanderthals.

Many instances of adaptive introgression—those, for example, that involve biochemical changes that do not affect appearance—are cryptic and were effectively undetectable before the development of modern molecular research methods. This is worth remembering when we look at the fossil record: The majority of adaptive genetic events do not have noticeable skeletal signs. Some cases of adaptive introgression, though, have readily visible effects, as when genes that increased drought tolerance spread from Utah cliffrose to bitterbrush. The introgressed bitterbrush looks more like cliffrose and can survive in places where ordinary bitterbrush cannot.[16] In this case, the population with introgressed genes reflects that introgression in its external appearance, but more often the effects of introgression are not readily apparent in the gross anatomy of an organism.

## BREEDING EXPERIMENTS

Applied geneticists are always conducting breeding experiments—usually for practical purposes in agriculture, often for research, sometimes for the sheer fun of it. In those experiments, they usually select for some trait (or for the absence of that trait): that is, they breed a new generation from those individuals with especially high (or low) values of a trait. The average value of that trait (or the absence of it) changes over generations and can eventually reach levels that differ greatly from the original population. If you doubt this, consider that the Chihuahua is the product of selection upon wolves. Change gradually slows (at least in small populations) and the trait plateaus. Sometimes this is because further change is physically impossible, but more often it is because the genetic variety of the population has become exhausted. When several different populations (drawn from the same base population) have undergone such selection and plateaued, sometimes the breeder will take the two best lines and cross them. Some such efforts are unproductive, but others succeed in producing a population with significantly higher trait values.

Life is a breeding experiment, of course. And it looks as if the African-Neanderthal cross has worked out pretty well—so far, anyway.

The key point here is that it would take only a very limited amount of interbreeding for modern humans to have picked up almost every Neanderthal allele with any significant advantage. Limited interbreeding would mean that neutral genes in humans today would look almost entirely African—which they do—while at the same time we might carry a number of functional alleles that originated in Neanderthals. Those alleles would be ones that mattered, ones that made a difference.

This raises the question of just what the Neanderthals might have had to offer. The popular impression is that they were backward, almost bestial—and it's certainly true that moderns had capabilities that Neanderthals lacked. But in archaeological artifacts from as recently as 100,000 years ago, it's hard to see any real differences in the material culture of Neanderthals and the material culture of Africans—so the Neanderthals can't have been all that far behind.

The alleles most obviously worth stealing would be those that implemented adaptations to local conditions in Europe.

That might mean, for example, acquiring the ability to tolerate cold weather, resist local diseases, or adjust to big swings in the length of the day over the course of a year (in contrast to the tropics, where the length of the day does not vary much). These kinds of adaptations, along with the more sophisticated, technological solutions to cold characteristic of modern humans, such as building shelters and so on, may have been important in human settlement of the far north, and eventually of the Americas.

These sorts of changes were important—adaptation to local climates and pathogens was obviously necessary for humans to succeed in northern climates. But on the whole, they are not all that interesting. Obviously, even penguins are better adapted to cold than humans. If that is all the Neanderthals had to offer, the question of interbreeding with them would not matter so much. The most interesting genetic changes are surely those that change minds rather than bodies. And there are several lines of argument that suggest that the Neanderthals may have had something to contribute along those lines as well.

## Changing Minds

Neanderthals had developed larger brains during their time in Europe, just as modern human ancestors had in Africa (and to some extent as the archaic populations in Asia had as well). Those large brains paid off in increased fitness in both populations, else they'd never have come into existence, but there may have been functional differences.

There were deep differences between *Homo sapiens* and *Homo neanderthalensis* in way of life, with Neanderthals being high-risk, highly cooperative hunters, rather like wolves, while anatomically modern humans in Africa probably had a mixed diet and were more like modern hunter-gatherers. Those differences could mean that big Neanderthal brains were solving different sorts of problems than big African brains. As a purely hypothetical example, Neanderthals, facing high risks as ambush hunters of big game, might have benefited from an ability to imagine and anticipate the reactions of their prey—call it "theory of animal mind." Neanderthals were strong and heavily built, but their hunting success depended on brainpower to a far higher degree than that of lions or wolves. Their intelligence made their way of life possible via the use of tools and weapons, but there must have been other ways in which their big brains aided survival. Improved accuracy in guessing just how a wounded bison would react could well have kept a Neanderthal from having his ribs kicked in, for example. There's a precedent for this notion of one species having a theory of mind in dealing with another species: Wolves can't take a hint, but dogs have an evolved ability to read people.[17]

And yet, European Neanderthals probably faced many of the same life problems that African humans did. To some degree, big brains may have been solving the same problems in both populations. Even if that is the case, though, we can be certain that those problems were not solved in exactly the same way. Examples demonstrating how natural selection works can shed light on this concept. Let's look again at human adaptation to malaria. We see hemoglobin variants in both Africa and Southeast Asia (sickle cells and Hemoglobin E [HbE]), but they're not the *same* variants. Although both alleles protect against malaria, there's no reason to assume that one works exactly like the other (or even as well as the other). HbE, for example, definitely has fewer negative side effects than sickle cells.

We see a similar pattern in human adaptation to high altitude. On one hand, the Amerindians of the high Andes have barrel chests and blood crammed with red cells; the Tibetans, on the other hand, have much lower levels of hemoglobin but breathe faster to take in more oxygen. Both peoples are far better adapted to high altitude than flatlanders, but the Tibetan adaptation is apparently more effective, since their babies are plumper and healthier. Adaptation depends on the supply of favorable mutations, which are generated randomly; thus, two populations facing the same problem may well find different solutions, and those solutions need not be equally efficient. Neanderthals and the anatomically modern humans of Africa faced some of the same conditions and adapted to them, but they did not necessarily do so in the same way or with the same degree of efficiency.

As we mention elsewhere in this work, sometimes variation in human personality is best explained as a genetically based alternative behavioral strategy that works well when rare but whose advantage dissipates as bearers of that strategy become more common. For example, many suspect that human sociopaths, individuals who are well-designed "cheaters," like con artists, can prosper when they are rare but suffer fitness loss as they become more common and others become more aware of them.[18]

There are many possible alternative strategies, and it is possible that the Neanderthals had some that never came into existence among the modern humans in Africa—and yet could succeed among them, particularly since increased innovation had shaken up society. So, it could be the case that as the modern humans moved north and came into contact with Neanderthals, they picked up alternative strategies for solving various problems—strategies with a genetic basis that came to them not by observation but through introgression and natural selection, and which depended upon new mental functions and cognitive processes.

### Paths on Fitness Landscapes

Another point: Ongoing natural selection in two populations can allow evolutionary events to occur that would be impossible in a single well-mixed population, since it allows for simultaneous exploration of divergent paths. Natural selection is shortsighted: Alleles increase in frequency because of their current advantage, not because they might someday be useful. Think of various possible solutions of some problem as hills, with higher hills corresponding to better solutions. Natural selection climbs up the first hill it chances upon; it can't see that another solution has greater possibilities in the long run. Not only that: Since the environmental conditions of Europe and Africa were significantly different, evolution could try solutions in Europe that couldn't be explored in Africa, because the initial step along that path had negative payoffs in Africa. In Europe, for example, you had to worry about staying warm enough, whereas Africans faced heat stress: These issues were important considerations in the evolution of larger brains. It may be that the relative unimportance of heat stress in Europe opened up some evolutionary pathways that had greater long-term possibilities than the ones that developed in Africa.

Consider an analogy from the history of technology. Somewhere back in late classical times, the use of the camel was perfected—a better saddle was developed, for example, one that allowed camels to carry heavy loads efficiently. Throughout most of the Middle East and North Africa, camels were (after those developments) a superior means of land transportation: They were cheaper than ox-drawn wagons and not dependent upon roads. Over a few centuries, people in areas where camels were available abandoned wheeled vehicles and roads almost entirely.[19] You can still see the effects in the oldest sections of some cities in the Arab world, where the alleys are far too narrow to have ever passed a cart or wagon. Europeans, not having camels, had to stick with wheeled vehicles, which were clearly more expensive, given the infrastructure they required. But as it turned out, wheeled vehicles—in fact, the whole road/wheeled vehicle system—could be improved. Back then, when camels seemed so much better, who knew that someday there would be horse collars and nailed horseshoes, then improved bridge construction, suspensions that reduced road shock, macadamized roads, steam power, internal combustion engines, and ultimately the nuclear Delorean. The motto here is that sometimes the apparently inferior choice has a better upgrade path: Evolution can't know this, and we aren't particularly good at recognizing it ourselves. On the genetic level, it translates as follows: Natural selection may solve the same problems differently in different populations, and what appears to be the most elegant solution at the time may not in fact turn out to be the one that works best in the long run. The seemingly inferior choice may come out on top down the road. It is easy to think of plausible cases: Imagine, for example, that excess heat production limited the trend toward larger brains in Africa, while in the climate of Europe heat was not much of a problem. Later, as evolution fine-tuned the physiology of large brains, much of the heat problem was solved—and so the new brain could then spread in Africa as well.

### Spare Genetic Diversity for the Future

Lastly, we should consider that even a slight degree of Neanderthal admixture would have increased the amount of genetic variation among modern humans, and those imported alleles could have been useful in solving future adaptive problems even if they were not particularly advantageous back in the Upper Paleolithic. We now know that the transition to agriculture posed many challenges that led to strong selective pressures on farmers. Most of the adaptive responses to agriculture were probably the result of new mutations, but some must have made use of preexisting genetic variation, which would have included any alleles that we picked up from Neanderthals or other archaic humans. Think of it this way: Modern humans and Neanderthals were both unsuited to agriculture and civilization, so Africans were not much more likely to carry alleles preadapted to agriculture than Neanderthals were. The solutions to agricultural challenges could have come from either camp.

We're not saying that Neanderthals were competitively superior: After all, we're here, and they're not. But it is highly likely that, out of some 20,000 genes, at least a few of theirs were worth having.

Not only was interbreeding between Neanderthals and moderns likely and potentially important, there is evidence indicating that it actually occurred. That evidence is of two kinds, skeletal and genetic.

## SKELETAL EVIDENCE

Neanderthal anatomy differed in a number of ways from that of anatomically modern humans, as we have noted before. There are particular details that are relevant to interbreeding. One is the *occipital bun*, a bulge at the back of the skull, which was very common among Neanderthals but is rare among people today. Another is the retromolar space, a gap between the last molar and the back of the mouth. These and other skeletal details characteristic of Neanderthals were unusually common among the modern humans who were their immediate successors but have declined in frequency over time.[20] The skeletons we've found of modern humans during the Upper Paleolithic don't have the pulled-forward face that is so distinctive of Neanderthals—which is consistent with the idea that there wasn't a *lot* of gene flow. Complex craniofacial features probably depend on many genes working together, so such features are unlikely to show up if Neanderthal genes are uncommon in modern humans of the period. There have been claims that certain early but clearly modern human skeletons have several distinctive Neanderthal skeletal features, however, which could indicate recent admixture.[21]

We think that the skeletal evidence suggests that there was significant Neanderthal admixture, but we also recognize that this evidence is not by itself definitive. Considering the possibility of convergent evolution, the situation is complex. One problem is that skeletal features, like almost everything else, evolved for a reason: They somehow increased fitness for the Neanderthals in their environment. It is therefore possible that some features similar to those of the Neanderthals evolved independently in Cro-Magnons (that is, in anatomically modern humans of the Upper Paleolithic) because they fulfilled the same functions. Moreover, only Neanderthal traits that were adaptive are likely to have introgressed and reached significant frequency today. Fortunately we're not limited to skeletal evidence—we are rapidly acquiring genetic evidence that bears on this question.

## GENETIC EVIDENCE

The first investigations of modern humans attempting to detect remnants of ancient lineages looked at mitochondrial DNA and Y chromosomes. Both are of interest because they are inherited from just one parent (the Y chromosome from the father, the mtDNA from the mother) and because they do not recombine. Extensive sampling has shown no evidence of variants that might have existed in archaic human populations such as Neanderthals.[22] The data, in other words, are consistent with low (or zero) gene flow from archaic to modern populations. This pattern might also have arisen if Neanderthal mtDNA and Y chromosomes didn't mesh well with the genetic background of anatomically modern humans and reduced fitness in some way. In that case, they would have dwindled with time and could be rare or nonexistent today even if they had once been moderately common in modern humans.

A number of recent reports, however, provide evidence that people do retain some autosomal alleles from archaic humans.[23] Some of these reports have detected odd patterns in our genome as a whole, whereas others have looked closely at particular unusual genes.

V. Plagnol and J. D. Wall found that the pattern of linkage disequilibrium—that is, of the history of chromosomes having broken and reformed—among SNPs (single nucleotide polymorphisms, or single base differences between chromosomes) in the human genome was inconsistent with an unstructured ancient population, estimating that 5 percent of genetic variation among both Europeans and sub-Saharan Africans originated in archaic humans such as the Neanderthals.[24] This is interesting, in that evidence for introgression is nearly as strong among Africans as it is among Europeans. This is what one would expect to happen if many of the alleles picked up from Neanderthals or eastern archaic humans were generally advantageous and spread very widely. There may also have been significant archaic populations somewhere in Africa: There are some apparently archaic variants found only in Pygmies, which suggests an African origin. Conditions for fossilization are poor in much of Africa west of the Rift (for example, chimpanzees have almost no fossil record)—and there may well have been hominid populations other than anatomically modern humans in the blank spots in those Africa maps. Since some of these alleles are found at high frequencies in people today, and since the overall level of admixture was probably low, they probably gave a fitness advantage—in other words, were adaptive.

P. D. Evans and his colleagues at the University of Chicago looked at microcephalin (MCPH1), a very unusual gene that regulates brain size.[25] They found that most people today carry a version that is quite uniform, suggesting that it originated recently. At the same time, it is very different from other, more varied versions found at the same locus in humans today, all of which have many single-nucleotide differences among them. More than that, when there are several different versions of a gene at some locus, we normally find some intermediate versions created by recombination, that is, by chromosomes occasionally breaking and recombining. In the case of the unusual gene (called *D* for "derived") at the microcephalin locus, such recombinants are very rare: It is as if the common, highly uniform version of microcephalin simply hasn't been in the human race all that long in spite of the high frequency of the new version in many human populations. The researchers estimated that it appeared about 37,000 years ago (plus or minus a few tens of thousands of years). And if it did show up then, Neanderthals are a reasonable, indeed likely, source.

Another interesting possibility involves FOXP2, a gene that plays a role in speech that was replaced by a new variant some 42,000 years ago.[26] This is very recent in evolutionary terms, and there is evidence that the same version of that gene existed in Neanderthals.[27] If the new FOXP2 allele is really that recent in modern humans, it is likely that the migrating humans picked it up from Neanderthals, since that's about the time they encountered them in their expansion out of Africa. The idea that we might have acquired some of our speech capabilities from Neanderthals may be surprising, but it is not impossible. The timing of the acquisition is certainly consistent with the creative explosion. If it is true that we gained the gene by means of introgression, then the version of FOXP2 in the Neanderthals is likely to be older and have more variation than it does in modern humans. Further sequencing efforts on the skeletal remains of Neanderthals should eventually confirm or refute this possibility.

If FOXP2 is indeed a "language gene" and responsible, perhaps, for some of the creative explosion of modern humans in Europe and northern Asia, it would explain a major puzzle about modern human origins. There were at least two streams out of Africa 50,000 years ago, one northward into Europe and central Asia, and another eastward around the Indian Ocean to Australia, New Guinea, and parts of Oceania. There is no trace of any creative explosion in populations derived from the southern Indian Ocean movement, who brought and retained Neanderthal-grade technology and culture.[28]

## CONCLUSION

A burst of innovation followed the expansion of modern humans out of Africa. Signs of that change existed in Africa before the expansion, but the pattern became much stronger in Europe some 20,000 years later, after anatomically modern humans had encountered and displaced the Neanderthals. That transition to full behavioral modernity—as seen in the archaeological record—occurred patchily and finished later in other parts of Eurasia. We argue that even limited gene flow from Neanderthals (and perhaps other archaic humans) would have allowed anatomically modern humans to acquire most of their favorable alleles. We believe that this sudden influx of adaptive alleles contributed to the growth of the capabilities that made up the "human revolution," and we believe that this introgression from archaic human populations will prove central to the story of modern human origins.

So by 40,000 years ago, humans had become both anatomically and behaviorally modern (which is not to say they were exactly like people today). They had vastly greater powers of innovation than their ancestors, likely owing in part to genes stolen from their Neanderthal cousins. The speed of cultural change increased by tens of times, and when the glaciers retreated and new opportunities arose, it accelerated further.

# 3

# AGRICULTURE: THE BIG CHANGE

Favorable mutations are rare, and many of those that do occur are lost by chance. In the small human populations of the Old Stone Age, establishing such mutations typically took hundreds of thousands of years. It's not that it took that long for favorable mutations to spread—the problem was generating them in the first place.

But as human population sizes increased, particularly with the advent of agriculture, favorable mutations occurred more and more often. Sixty thousand years ago, before the expansion out of Africa, there were something like a quarter of a million modern humans. By the Bronze Age, 3,000 years ago, that number was roughly 60 million. Favorable mutations that had previously occurred every 100,000 years or so were now showing up every 400 years.

One might think that it would take much longer for a favorable mutation to spread through such a large population than it would for one to spread through a population as small as the one that existed in the Old Stone Age. But since the frequency of an advantageous allele increases exponentially with time in a well-mixed population, rather like the flu, it takes only twice as long to spread through a population of 100 million as it does to spread through a population of 10,000.

Agriculture imposed a new way of life (new diets, new diseases, new societies, new benefits to long-term planning) to which humans, with their long history as foragers, were poorly adapted. At the same time it led to a vast population expansion that greatly increased the production of adaptive mutations.[1] So agriculture created many new problems, but it created even more new solutions. Earlier innovations had also helped to increase population size and thus had speeded up human evolution, but agriculture had a far greater effect and is in a class of its own.

Naturally, increased population size had a similar impact on the generation of new ideas. All else equal, a large population will produce many more new ideas than a small population, and new ideas can spread rapidly even in large populations. In *Guns, Germs, and Steel*, Jared Diamond observed: "A larger area or population means more potential inventors, more competing societies, more innovations available to adopt—and more pressure to adopt and retain innovations, because societies failing to do well will be eliminated by competing societies."[2] We take this observation a step further: There are also more *genetic* innovations in that larger population.

This is a new picture of recent human evolution. It implies that humans have changed not just culturally, but genetically, over the course of recorded history, and that we must allow for such changes when we try to understand historical events. The implications of this contention are vast: If correct, it means that peoples in different parts of the world have changed in varying ways, since they adopted different forms of agriculture at different times—or in some cases not at all.

Since genetic change wasn't uniform, discrete populations came to differ genetically from one another, and sometimes those genetic differences conferred competitive advantages. We believe that such genetic advantages have played a role in migrations and population expansions—and thus are important in explaining the current distribution of languages and peoples. In fact, history looks more and more like a science fiction novel in which mutants repeatedly arise and displace normal humans—sometimes quietly, simply by surviving, sometimes as a conquering horde.

It's probable that the evolutionary response to farming also affected the distribution of cognitive and personality traits, and that these changes played a crucial role in the development of civilization and the birth of the scientific and industrial revolutions.

## SETTING THE STAGE

When the Ice Age ended around 10,000 BC, the world became warmer and wetter, and the climate became more stable. Carbon dioxide levels increased, which increased plant productivity. The stage for agriculture was now set—and this time the actors were ready as well.

Although there had been other interglacial periods in the past, early humans had never developed agriculture then. We suspect that increases in intelligence made agriculture possible, but the route may have been indirect. For example, the invention of better weapons and hunting techniques, combined with other technologies that let humans make better use of plant foods, could have led to lower numbers, or even extinction, of key game animals—which would have eliminated an attractive alternative to farming.

Farming appeared first in the Fertile Crescent of Southwest Asia. By 9500 BC, we see the first signs of domesticated plants: first wheat and barley, then legumes such as peas and lentils.[3] From there farming spread in all directions, showing up in Egypt and western India by 7000 BC and gradually moving into Europe and India. Around 7000 BC, rice and foxtail millet were domesticated in China. Animals were domesticated on a similar timeline, with the Middle East in the lead. Goats were tamed around 10,000 BC in Iran, sheep about 1,000 years later in Iraq. Both the taurine cattle we're familiar with in the Middle East and the humped zebu cattle in India were domesticated around 6000 BC.

Agriculture came later to the rest of the world. In some cases it spread by a geographic expansion of farmers, in others through hunter-gatherers adopting already-existing methods of agriculture, and in yet others by hunter-gatherers independently inventing their own forms of agriculture. In Europe, agriculture was spread by Middle Eastern immigrants and by native Europeans learning to grow Middle Eastern crops such as wheat and barley. In sub-Saharan Africa, geographic barriers and climatic differences blocked adoption of most Middle Eastern crops and domesticated animals. There, agriculture appeared around 2000 BC and was based on locally domesticated crops such as sorghum and yams. The story is similar in the Americas, where the Amerindians were almost entirely cut off from the rest of the world and had to domesticate their own crops. (Some of those, such as maize and potatoes, are among the most important crops in the world today.)

Agriculture comprised what was surely the most important set of innovations since the expansion of modern humans out of Africa, resulting in changes in human diet, disease exposure, and social structure. Another consequence (one of great evolutionary significance) was a huge population boom. Human numbers had already been on the increase since the advent of behavioral modernity, partly as the result of migration into the far northern regions of Asia, over the sea into Australia, and across a land bridge into the Americas—all places that archaic humans had been unable to settle—and partly because of improvements in food production technology (such as nets and bows). An educated guess puts the total population of the world 100,000 years ago at half a million, counting both anatomically modern humans in Africa and archaic humans (Neanderthals and evolved erectus) in Eurasia. By the end of the Ice Age some 12,000 years ago, there may have been as many as 6 million modern humans—still hunter-gatherers, but far more sophisticated and effective hunter-gatherers than ever before.

Farming, which produces 10 to 100 times more calories per acre than foraging, carried this trend further. Over the period from 10,000 BC to AD 1, the world population increased approximately a hundredfold (estimates range from 40 to 170 times). That growth in itself transformed society—sometimes, quantity has a quality all its own. And as we have pointed out, this larger population was itself an important factor in evolution.

The advent of agriculture changed life in many ways, not all of them obvious. It vastly increased food production, but the nutritional quality of the food was worse than it had been among hunter-gatherers. It did not materially increase the average standard of living for long, since population growth easily caught up with improvements in food production. Moreover, higher population density, permanent settlements, and close association with domesticated animals greatly increased the prevalence of infectious disease.

The sedentary lifestyle of farming allowed a vast elaboration of material culture. Food, shelter, and artifacts no longer had to be portable. Births could be spaced closer together, since mothers didn't have to continually carry small children. Food was now storable, unlike the typical products of foraging, and storable food could be stolen. For the first time, humans could begin to accumulate wealth. This allowed for nonproductive elites, which had been impossible among hunter-gatherers. We emphasize that these elites were not formed in response to some societal need: They took over because they could.

Combined with sedentism, these developments eventually led to the birth of governments, which limited local violence. Presumably, governments did this because it let them extract more resources from their subjects, the same reason that farmers castrate bulls. Since societies were generally Malthusian, with population growth limited by decreasing agriculture production per person at higher human density, limits on interpersonal violence ultimately led to a situation in which a higher fraction of the population died of infectious disease or starvation.

All these changes generated new selective pressures, which is another way of saying that humans didn't fit the new environment they had created for themselves, so the species was under pressure to adapt. Because of the newness of the environment, genetic improvements were relatively easy to find—definitely easier at this point than finding ways to become better hunter-gatherers. Modern humans had been adapting to their hunting-gathering lifestyle for a very long time and had already exhausted most such possibilities. Adaptation to the farming life was doable, but as always, it would require concrete genetic changes.

## GENETIC RESPONSE

When agriculture was new, natural selection must have operated with the genetic variation that already existed, just as it does in small-scale artificial-selection experiments. Such experiments cause changes in the frequency of existing alleles.

Most preexisting genetic variation must have taken the form of a few neutral variants of each gene—variants that are not significantly different from each other. They may well do something, but the neutral alleles all do the same thing. We doubt if many of those neutral genes turned out to be the solution for the problems faced by the future farmers of Eurasia. More likely, preexisting functional variation mattered more. For example, there is a gene whose ancestral form helps people to conserve salt. Since humans spent most of their history in hot climates, this variant was generally useful. A high frequency of this ancestral allele among African Americans probably plays a role in their increased risk of high blood pressure today. In tropical Africa, in fact, almost everyone has the ancestral version of the gene. In Eurasia, a null variant (one that does nothing at all) becomes more and more common as one moves north.[4] Perhaps the gene's action of promoting salt conservation becomes harmful—by causing higher blood pressure—in cooler areas, where people sweat less and lose less salt.

Significantly, the null allele is the same in both Europe and eastern Asia—which suggests that it originated in Africa and is ancient. If it had separate European and Asian origins, then we would expect to see different versions in the two regions, just as different broken pigment genes lead to light skin in the two regions.

The most reasonable explanation for this dud salt-conservation gene is that parts of Africa (before the expansion out of Africa) were cool enough that salt retention was not a major concern, so that in these regions an inactive form of the gene was in fact advantageous. This might have happened in Ethiopia during glacial periods, considering that the climate on the Ethiopian plateau is moderate even today. If so, the null allele would represent preexisting adaptive variation caused by environmental variations inside Africa rather than neutral variation. Such internal variation inside Africa must have helped prepare humans for environments outside Africa.

Another kind of preexisting genetic variation would have consisted of *balanced polymorphisms*. Balanced polymorphisms occur within a population when the population maintains two different alleles of a gene, and the reason the polymorphism can be stable is that heterozygous individuals will have greater fitness than homozygous individuals. A *heterozygote advantage* exists, for example, in sickle cell and other malaria defenses. There are also alleles that have positive effects when rare, but whose advantages decrease as they become common, eventually becoming negative (this is called *frequency-dependent* selection). Some of the most interesting examples involve behavior and lend themselves to a game-theory analysis.

The best-known model is the hawk-dove game, where some individuals are genetically aggressive while others are genetically peaceful. When hawks are rare, they easily defeat doves and have higher fitness. As they become more common, however, they run into other hawks more often and have costly fights that decrease their fitness. At some frequency, the fitness of hawks and doves is the same, leading to a balanced polymorphism.[5]

Balanced behavioral polymorphisms could respond quickly to new selective pressures. If the original mix was 50 percent doves and 50 percent hawks, an environmental change that raised the costs of aggressive behavior would lead to a shift in frequency—say to 70 percent doves and 30 percent hawks. This kind of evolutionary change is very rapid, especially when compared to new sweeping genes, which are rare in the beginning and take thousands of years to reach frequencies of 20 percent or more. If the doves acquired a selective advantage of 5 percent, that change (from 50 percent to 70 percent) could occur in less than ten generations.

Human genetic variation was limited in the days before agriculture, in part because populations were small, and it was often not useful, since many of the changes that were favored among agriculturalists would actually have been deleterious among their hunter-gatherer ancestors. This means that some of the alleles with the right effects in farmers would have been extremely rare or nonexistent in their hunter-gatherer ancestors. For example, variants of G6PD (for glucose-6-phosphate dehydrogenase) with reduced function protect against falciparum malaria but also have negative effects, especially in men. Today, those G6PD variants have a net positive effect in malarious regions and have become common in many populations. Before the spread of falciparum malaria, those variants likely had a net negative effect in all populations, and so were extremely rare.

Therefore, new mutations must have played a major role in the evolutionary response to agriculture—and as luck would have it, there was a vast increase in the supply of those mutations just around this time because of the population increase associated with agriculture. We're not saying that the advent of agriculture somehow called forth mutations from the vasty deep that fitted people to the new order of things. Mutations are random, and as always, the overwhelming majority of them had neutral or negative effects. But more mutations occurred in large populations, some of them beneficial. Increased population size increased the supply of beneficial mutations just as buying many lottery tickets increases your chance of winning the prize.

By the beginnings of recorded history some 5,000 years ago, new adaptive mutations were coming into existence at a tremendous rate, roughly 100 times more rapidly than in the Pleistocene. This means that recent human evolution differs qualitatively from typical artificial selection acting on domesticated animals. It is simply a matter of scale. In the artificial-selection experiments, which typically involve no more than tens or hundreds of animals, very few new favorable mutations occur, and selection must act primarily on preexisting genetic variation. In recent human evolution, we're talking anywhere from millions to hundreds of millions of individuals, all of them potential mutants, so most of the advantageous variants would have been new.

You might think that alleles that were already common would be more likely than new variants to grow to high frequency under agriculture. It stands to reason that the new mutations, which would start out with a single copy, would face disadvantages. But that reasoning underestimates the effect of the advantage that the mutation conferred on the individual who carried it and his or her descendants. Even a single copy of an advantageous gene has a fair chance of succeeding (10 percent for a gene with a 5 percent advantage), and exponential growth allows it to spread rapidly. Many new mutations must have occurred in those large farming populations, and the great majority of the sweeping genes must have been new.

Not only did post-agricultural evolution involve much higher numbers than would be possible in any artificial-selection experiment, it also involved a much longer time frame. Post-agricultural evolution occurred over some 400 generations, which would be impractical for selection experiments using mammals. That long time scale also makes for a qualitative difference, since it is long enough to allow new mutations to rise to high frequency and make up a major part of adaptive variation.

Recent studies have found hundreds of ongoing sweeps—sweeps begun thousands of years ago that are still in progress today. Some alleles have gone to fixation, more have intermediate frequencies, and most are regional. Many are very recent: The rate of origination peaks at about 5,000 years ago in the European and Chinese samples, and at about 8,500 years ago in the African sample. There are so many sweeps under way, in fact, that we can do some useful statistical analysis. Often we have some idea of a gene's function—for example, by seeing what tissues it is highly expressed in, or by knowing what goes wrong when it's inactivated. Using that information, we can look at the hundreds of genes undergoing sweeps and see what kinds of jobs they do. And when we do that kind of analysis, we see that most of the sweeping alleles fall into a few functional categories: Many involve changes in metabolism and digestion, in defenses against infectious disease, in reproduction, in DNA repair, or in the central nervous system.

## YOU ARE WHAT YOU EAT

Early farmers ate foods that hunter-gatherers did not eat, or at least they ate them in much greater quantities, and at first they were not well adjusted to the new diet. In Europe and western Asia, cereals became the dietary mainstay, usually wheat or barley, while millet and rice became the primary foods in eastern Asia. Those early farmers raised other crops, such as peas and beans, and they ate some meat, mostly from domesticated animals, but it looks as if the carbohydrate fraction of their diet almost tripled, while the amount of protein tanked.[6] Protein quality decreased as well, since plant foods contained an undesirable mix of amino acids, the chemical building blocks of which proteins are made. Almost any kind of meat has the right mix, but plants often do not—and trying to build muscle with the wrong mix is a lot like playing Scrabble with more Q's than U's.

Shortages of vitamins are also likely to have been a problem among those early farmers, since the new diet included little fresh meat and was primarily based on a very limited set of crops. Hunter-gatherers would rarely have suffered vitamin-deficiency diseases such as beri-beri, pellagra, rickets, or scurvy, but farmers sometimes did. There is every reason to think that early farmers developed serious health problems from this low-protein, vitamin-short, high-carbohydrate diet. Infant mortality increased, and the poor diet was likely one of the causes. You can see the mismatch between the genes and the environment in the skeletal evidence. Humans who adopted agriculture shrank: Average height dropped by almost five inches.[7]

There are numerous signs of pathology in the bones of early agriculturalists. In the Americas, the introduction of maize led to widespread tooth decay and anemia due to iron deficiency, since maize is low in bioavailable iron. This story is not new: Many researchers have written about the health problems stemming from the advent of agriculture.[8] Our point is that, over millennia, populations *responded* to these new pressures. People who had genetic variants that helped them deal with the new diet had more surviving children, and those variants spread: Farmers began to adapt to an agricultural diet. Humanity changed.

We are beginning to understand some of the genetic details of these dietary adaptations, which took several forms. Some of the selected alleles appear to have increased efficiency—that is to say, their bearers were able to extract more nutrients from an agricultural diet. The most dramatic examples are mutations that allow adults to digest lactose, the main sugar in milk. Hunter-gatherers, and mammals in general, stop making lactase (the enzyme that digests lactose) in childhood. Since mother's milk was the only lactose-containing "food" available to humans in days of yore, there wasn't much point in older children or adults making lactase—and shutting down production may have decreased destructive forms of sibling rivalry. But after the domestication of cattle, milk was available and potentially valuable to people of all ages, if only they could digest it. A mutation that caused continued production of lactase originated some 8,000 years ago and has spread widely among Europeans, reaching frequencies of over 95 percent in Denmark and Sweden. Other mutations with a similar effect have become common (despite starting several thousand years later) in some of the cattle-raising tribes in East Africa, so that 90 percent of the Tutsi are lactose tolerant today. These mutations spread quite rapidly and must have been very advantageous.

When you think about it, the whole process is rather strange: Northern Europeans and some sub-Saharan Africans have become "mampires," mutants that live off the milk of another species. We think lactose-tolerance mutations played an important role in history, a subject we will treat at some length in Chapter 6.

Some genetic changes may have helped to compensate for shortages in the new diet. For example, we see changes in genes affecting transport of vitamins into cells.[9] Similarly, vitamin D shortages in the new diet may have driven the evolution of light skin in Europe and northern Asia. Vitamin D is produced by ultraviolet radiation from the sun acting on our skin—an odd, plantlike way of going about things. Less is therefore produced in areas far from the equator, where UV flux is low. Since there is plenty of vitamin D in fresh meat, hunter-gatherers in Europe may not have suffered from vitamin D shortages and thus may have been able to get by with fairly dark skin. In fact, this must have been the case, since several of the major mutations causing light skin color appear to have originated after the birth of agriculture. Vitamin D was not abundant in the new cereal-based diet, and any resulting shortages would have been serious, since they could lead to bone malformations (rickets), decreased resistance to infectious diseases, and even cancer. This may be why natural selection favored mutations causing light skin, which allowed for adequate vitamin D synthesis in regions with little ultraviolet radiation.

There were other changes that ameliorated nasty side effects of the new unbalanced diets. The big increase in carbohydrates, especially carbohydrates that are rapidly broken down in digestion, interfered with the control of blood sugar and appears to have caused metabolic problems such as diabetes. A high-carbohydrate diet also apparently causes acne and tooth decay, both of which are rare among hunter-gatherers. More exactly, both are caused by infectious organisms, but those organisms only cause trouble in the presence of a high-carbohydrate diet.

Some of the protective changes took the form of new versions of genes involved in insulin regulation. Researchers in Iceland have found that new variants of a gene regulating blood sugar protect against diabetes.[10] Those variants have different ages in the three populations studied (Europeans, Asians, and sub-Saharan Africans), and in each population the protective variant is roughly as old as agriculture. Alcoholic drinks, also part of the new diet, had plenty of bad side effects, and in East Asia there are strongly selected alleles that are known to materially reduce the risk of alcoholism.

Clearly, the evolutionary responses to an agricultural diet must differ, since different peoples adopted different kinds of agriculture at different times and in different environments. This variation has caused biological differences in the metabolic responses to an agricultural diet that persist today, but it has also generated differences in every other kind of adaptive response to the new society. Agriculture began in the Middle East 10,000 years ago and took almost 5,000 years to spread throughout Europe. Amerindians in the Illinois and Ohio river valleys adopted maize agriculture only 1,000 years ago, but the Australian Aborigines never domesticated plants at all. Peoples who have farmed since shortly after the end of the Ice Age (such as the inhabitants of the Middle East) must have adapted most thoroughly to agriculture. In areas where agriculture is younger, such as Europe or China, we'd expect to see fewer adaptive changes—except to the extent that the inhabitants were able to pick up genes from older farming peoples. And we'd expect to see fewer adaptive changes still among the Amerindians and sub-Saharan Africans, who had farmed for even shorter times and were genetically isolated from older civilizations by geographical barriers. In groups that had remained foragers, there would presumably be no such adaptive changes—most certainly not in isolated forager populations.

Populations that have never farmed or that haven't farmed for long, such as the Australian Aborigines and many Amerindians, have characteristic health problems today when exposed to Western diets. The most severe such problem currently is a high incidence of type 2 diabetes. Low physical activity certainly contributes to that problem today, but genetic vulnerability is a big part of the story: Navajo couch potatoes are far more likely to get adult-onset diabetes than German or Chinese couch potatoes. The prevalence of diabetes among the Navajo is about two and a half times higher than it is in their European-descended neighbors, and about four times more common among Australian Aborigines than in other Australians. We think this is a consequence of a lesser degree of adaptation to high-carbohydrate diets. Interestingly, Polynesians are also prone to diabetes (with roughly three times European rates), even though they practiced agriculture, raising crops such as yams, taro, bananas, breadfruit, and sweet potato. We believe that their case still fits our general picture of incomplete adaptation, however. Among the Polynesians, adaptation would have been limited by the relatively small population size and the low rate of protective mutations it would have generated. In addition, settlement bottlenecks and limited contacts between the populations of the far-flung Polynesian islands would have interfered with the spread of any favorable mutations that did occur.

Our explanation of this susceptibility pattern differs from the well-known "thrifty genotype" hypothesis originally promulgated by James Neel. He suggested that pre-agricultural peoples were especially prone to famine and that the metabolic differences that led to diabetes in modern environments had helped people survive food shortages in the past.[11] This seems unlikely. The lower rungs of agricultural societies in Europe and East Asia usually suffered food shortages severe enough to cause below-replacement fertility, and weather-related crop failure often struck whole nations or even larger regions. Sometimes this led to famines severe enough to lead to widespread cannibalism, as seems to have occurred in the great famine that struck most of northern Europe from 1315 to 1317.

Hunter-gatherers should have been, if anything, less vulnerable to famine than farmers, since they did not depend on a small set of domesticated plant species (which might suffer from insect pests or fungal blights even in a year with good weather), and because local violence usually kept their populations well below the local carrying capacity.[12] State societies limited local violence, but in a Malthusian world, *something* always limits population growth. In this case, fewer deaths by violence meant more deaths due to starvation and infectious disease. Moreover, hunter-gatherer societies do not appear to have been divided into well-fed elites and hungry lower classes, a situation that virtually guarantees malnourishment and/or famine among a significant fraction of the population, whereas agricultural societies did have divisions of this sort. We believe that our explanation, based on the evolutionary response to a well-established increase in carbohydrate consumption among farmers, is more likely to be correct than an explanation based on the idea that hunter-gatherers were particularly prone to famine, a notion that has no factual support.

Most populations that are highly vulnerable to type 2 diabetes also have increased risks of alcoholism. This is no coincidence. It's not that the same biochemistry underlies both conditions, but that both stem from the same ultimate cause: limited previous exposure to agricultural diets, and thus limited adaptation to such diets.

Booze inevitably accompanies farming. People have been brewing alcoholic beverages since the earliest days of agriculture: Beer may date back more than 8,000 years. There's even a hypothesis that barley was first domesticated for use in brewing beer rather than bread. Essentially all agricultural peoples developed and routinely consumed some kind of alcoholic beverage. In those populations with long exposure, natural selection must have gradually increased the frequency of alleles that decreased the risk of alcoholism, due to its medical and social disadvantages. This process would have gone furthest in old agricultural societies and presumably would not have occurred at all among pure hunter-gatherers.

We must wonder why farming peoples didn't just evolve an aversion to alcohol. It seems as if that would have been a bad strategy, since moderate consumption of traditional, low-proof alcoholic drinks was almost certainly healthful. People who drank wine or beer avoided waterborne pathogens, which were a lethal threat in high-density populations. Alleles that reduced the risk of alcoholism therefore prevailed.

There is also some reason to believe that populations that have been drinking alcohol for hundreds of generations may have also evolved metabolic changes that reduced some of alcohol's other risks. In particular, we know that alcohol consumption by pregnant women can have devastating effects on their offspring. Those effects, called fetal alcohol syndrome, or FAS, include growth deficiency, facial abnormalities, and damage to the central nervous system. FAS is, however, far more common in some populations than in others: Its prevalence is almost thirty times higher in African American or Amerindian populations in the United States than it is among Europeans—even though the French, for example, have been known to take a drink or two. Some populations, such as those of sub-Saharan Africa and their diaspora, may run higher risks of suffering from FAS than others consuming similar amounts of alcohol. If so, study of the alleles protecting against FAS in resistant populations might lead to greater understanding of the biochemical mechanisms underlying the syndrome. With luck, we might be able to use that information to decrease the incidence of FAS in vulnerable populations.

This picture of adaptation to agricultural diets has two important implications: Populations today must vary in their degree of adaptation to such diets, depending on their historical experience, and populations must have changed over time.

For example, there must have been a time when no one was lactose tolerant, a later time in which the frequency was intermediate, and finally a time when it reached modern levels. In this instance, we have hard evidence of such change. In a 2007 study, researchers studied DNA from the skeletons of people who died between 7,000 and 8,000 years ago. These skeletons were from central and northern Europe, where today the frequency of the lactase-persistence variant is around 80 percent. None of those ancient northern Europeans had that allele.[13] In another study, a different group of researchers looked at central

European skeletons from the late Bronze Age, some 3,000 years ago. Back then, the gene frequency (judging from their sample) was apparently around 25 percent.[14] This shows that the frequency of lactose tolerance really has changed over time in the way indicated by the HapMap genetic data. The theory made sense, but experimental confirmation is always welcome. We expect that there will be many similar results (showing ongoing change in sweeping genes) in studies of ancient DNA over the next few years.

Over time, if our argument is correct, farming peoples should have become better adapted to their agricultural diets in many ways, and we might expect that some of the skeletal signs of physiological stress would have gradually decreased. Although such genetic adaptation clearly occurred, cultural changes that improved health must have occurred as well. For example, the adoption of new crops and new methods of food preparation would have improved the nutritional quality of the average peasant's diet. Of course, some of those new methods (polishing rice) and new crops (sugarcane)—actually made things worse. Adaptive change is slow and blind, but it is also sure and steady. Cultural change is less reliable.

But cultural change *is* important. Although many traditional archaeologists and anthropologists will probably see us as biological imperialists out to explain everything that ever happened with our pet genetic theories, we firmly believe that cultural change—new ideas, new techniques, new forms of social organization—were powerful influences on the historical process. We're simply saying that the complete historical analyst must consider genetic change as well as social, cultural, and political change. Once a list of battles and kings seemed plenty good enough, but life keeps getting more complicated.

# 4
# CONSEQUENCES OF AGRICULTURE

Agriculture reshaped human society, resulting in selective pressures that changed us in many ways. Some of those changes involved fairly obvious accommodations to new problems in nutrition and infectious disease. Others consisted of subtle psychological and cognitive changes, some of which eventually led to revolutionary social innovations—possibly including the birth of science. In this chapter, we discuss many of those evolutionary responses.

## INFECTIOUS DISEASE

Of course, diet is not the only thing that changed under agriculture. Farming revolutionized human infectious disease—but not in a good way.

The population expansion associated with farming increased crowding, while farming itself made people sedentary. Mountains of garbage and water supplies contaminated with human waste favored the spread of infectious disease. As farmers, humans acquired new commensals, animals that lived among them. We already had ride-along commensals such as lice and intestinal worms—now we had rats and mice as well, which spread devastating diseases such as typhus and bubonic plague.

Quantitative changes in population density and disease vectors resulted in qualitative changes in disease prevalence—not only did old infectious diseases become a more serious threat, entirely new ones appeared.

Most infectious diseases have a critical community size, a number and concentration of people below which they cannot persist. The classic example is measles, which typically infects children and remains infectious for about ten days, after which the patient has lifelong immunity. In order for measles to survive, the virus that causes it, the paramyxovirus, must continually find unexposed victims—more children. Measles can only persist in a large, dense population: Populations that are too small or too spread out (under half a million in close proximity) fail to produce unexposed children fast enough, so the virus dies out. This means that measles, at least in the form we know it today, could not have existed in the days before agriculture—there was no concentrated population that large anywhere on earth. (The virus that causes chicken pox is different: It lingers in the nervous system and often reemerges late in life in the form of shingles, which can be incredibly painful. Children can catch chicken pox from their grandparents—cycle of life! Since the critical community size of chicken pox is less than 100 people, epidemiologists judge that it has probably been around for a long time.)

In any case, the new conditions that accompanied agriculture brought more than measles. Many other diseases that just didn't exist in hunter-gathering days could now thrive as well. Some originated as mutated versions of milder infectious diseases that already existed in humans; we picked up others (probably most) from animals, especially domesticated herd animals. Later, as trade and travel increased, civilizations exchanged some of their regional diseases, with disastrous results.

Infectious disease was thus a far bigger threat to farmers than it had been to hunter-gatherers—which meant that farmers experienced strong selective pressures on that account. They eventually developed much more effective genetic defenses against infectious disease than those sported by their Neolithic ancestors, and these defenses were also far more effective than those possessed by people who remained hunter-gatherers.

The best-understood genetic defenses are those that protect people against falciparum malaria. There are several kinds of malaria, but falciparum is the most serious and accounts for the most deaths. Increased population density and the use of slash-and-burn agricultural techniques (cutting and burning forests to create fields) may have favored the spread of this virulent form of malaria. The trend was particularly unpleasant in Africa, where mosquitoes that preferred humans to animals evolved, facilitating transmission of this deadly disease.

Wherever falciparum malaria has existed for a long time, mainly in the tropical areas of the Old World, people have developed genetic defenses against it, and the side effects of those defenses account for most cases of genetic disease in populations originating in those regions. We know a lot about malaria defenses because they cause illness, and more time and effort has been spent on medical research than on understanding the evolutionary underpinnings of the disease. This is not surprising, since these illnesses have been so troubling in tropical areas. But understanding the root causes of the medical conditions may be worthwhile: A little more evolutionary thought in medicine might actually have practical payoffs.

The most important mutations that protect against malaria are those that change some feature in the red blood cells that are the primary target of the malaria parasite—usually, the hemoglobin molecule (for example, sickle cell hemoglobin [HbS], hemoglobin C [HbC], hemoglobin E [HbE], alpha- and beta-thalassemia, Melanesian ovalocytosis, and glucose-6-phosphate dehydrogenase [G6PD] deficiency). We also know of a number of alleles (such as the glycophorin C variant in New Guinea[1]) that are almost certainly malaria defenses but do not cause noticeable disease as side effects. In fact, it looks as if the well-known defenses, such as sickle cell, that cause obvious disease are only the tip of the iceberg.

The expensive malaria defenses (defenses with serious side effects) are far more common than any single genetic disease caused by random mutations. Some 400 million people, 7 percent of the world's population, have G6PD deficiency, which can be serious. About 250,000 children are born with sickle-cell anemia each year (which is *very* serious), while about 20,000 boys are born with Duchenne muscular dystrophy, one of the most common of all mutation-driven genetic diseases.[2]

These malaria defenses became common because they gave an advantage to carriers (people with one copy of the gene variant); however, they cause problems (from mild to lethal) in people with two copies. This is unusual: We seldom see such crude adaptations in other functions. For example, humans don't have an allele that makes carriers run faster while crippling those with two copies. Normally, genes work together in an efficient and coordinated way. We think that this evolutionary sloppiness exists because falciparum malaria, as we know it today, has not been around very long—perhaps as little as 4,000 years. The same appears to be true of the antimalaria genetic defenses. For example, the main African variety of G6PD deficiency is roughly 2,500 years old, HbE in Thailand is roughly 2,000 years old.[3]

These adaptations to falciparum malaria were both recent and local. They occurred in the tropical and subtropical areas of the Old World: Peoples who lived in the cooler parts of Eurasia, in Australia, and in the Americas either remained unexposed or were only exposed even more recently. Malaria reshaped the human genome, but only in some peoples. It has been one of the forces differentiating human populations over the past few thousand years.

Malaria defenses are only one example of a more widespread phenomenon. Recent whole-genome selection scans suggest that there have been many other genetic changes related to defense against disease. Again, the extent of these adaptations has varied regionally.

We see evidence of a number of cases in which new alleles related to pathogen defense and the immune system have rapidly reached high frequency: These alleles involve the production of antibodies, control of white cells that attack intruder organisms and infected cells, genes affecting viral infection, and cellular interaction with pathogens such as *Helicobacter pylori*, the bacterium that causes most stomach ulcers and stomach cancer. Again, most such changes are regional. But even before we began to discover these new defenses, it was obvious that something of the sort must exist, since some populations were much more vulnerable than others to diseases such as smallpox and influenza.

It's time to address the old chestnut that biological differences among human populations are "superficial," only skin-deep. It's not true: We're seeing genetically caused differences in all kinds of functions, and every such difference was important enough to cause a significant increase in fitness (number of offspring)—otherwise it wouldn't have reached high frequency in just a few millennia. These were not just superficial changes affecting things like hair color, skin color, and the shape of the nose, although even those apparently superficial differences sometimes had important consequences. Some of these differences were far from being superficial or insignificant and profoundly affected the populations in which they appeared, sometimes in unexpected ways. They had a major influence on history; some continue to shape the course of events today.

Populations that experienced different ecological histories had different evolutionary responses. In the case of infectious disease, it was in the main population centers of the Old World that human populations developed the strongest defenses. Populations isolated from the Old World diseases did not have an opportunity to develop such protections.

Amerindians, for example, experienced very little infectious disease. The story is similar in other isolated populations, such as the Australian Aborigines, Polynesians, and the inhabitants of the Andaman Islands: They didn't experience millennia of infectious disease, didn't evolve improved defenses as most Old Worlders did, and were decimated upon contact with the wider world.

## SKIN-DEEP

We now understand quite a bit about the genetic changes that led to light skin in northern Eurasians. At least we know what happened in Europe and Asia (China and Japan), the non-African populations studied in the HapMap.

In each of these populations, a number of genes have been replaced by—or are in the process of being replaced by—new variants that produce lighter skin color than the dark skin seen in sub-Saharan Africans, who mostly have the ancestral human alleles. Interestingly, the sets of changes driving light skin color in China are almost entirely different from those performing a similar function in Europe. In most cases the mutations involve changes in different genes, and even when the same gene is involved, usually the common mutations at the opposite ends of Eurasia are not the same. So in this example, as in lactase persistence and a number of other cases, it turns out that similar traits in different populations are the product of convergent evolution and are quite different at the level of biochemistry and genetics. Sometimes racial similarities are only skin-deep.

Many of these changes seem to be quite recent. The mutation that appears to have the greatest effect on skin color among Europeans and neighboring peoples, a variant of SLC24A5, has spread with astonishing speed. Linkage disequilibrium—that is, the degree to which the genome is surprisingly uniform around this gene—suggests that it came into existence about 5,800 years ago, but it has a frequency of about 99 percent throughout Europe and is found at significant levels in North Africa, East Africa, and as far east as India and Ceylon. If it is indeed that recent, it must have had a huge selective advantage, perhaps as high as 20 percent. It would have spread so rapidly that, over a long lifetime, a farmer could have noticed the change in appearance in his village. Again, if it is that recent, it must have had a more limited distribution in early historical times, particularly in peripheral areas: In fact, this may explain the Roman impression that the Picts of Scotland were dark-skinned.

As noted in Chapter 3, the recent sweeps of genes causing light skin might have been driven by an increased need for vitamin D among farmers living in high-latitude regions with low levels of ultraviolet radiation. But there are other possibilities. In the Old World tropics, such as sub-Saharan Africa, Melanesia, and New Guinea, the ancestral condition—dark skin—was favored by selection. Palefaces didn't prosper. But in higher-latitude regions, such as Europe and northern Asia, skin could be lighter. Many genes have more than one function: It may be that genes that produce dark skin pigments were now free to change in ways that enhanced some of their other functions, giving some kind of benefit other than increased vitamin D production.

We know of an example in fish that illustrates the same principle. In humans, OCA2 (for oculocutaneous albinism II) is a gene involved in the melanin pathway—if you have two broken copies, you're an albino. It also affects eye color: A particular variant that has increased rapidly in Europe is the main cause of blue eyes. Species of fish trapped in caves—this all relates, trust us—lose their eyesight and become albinos over many generations. But researchers have found that OCA2 is mutated in a number of different species of blind cave fish in Mexico, all descended from tetras. The mutations causing albinism in these fish are different from each other and originated independently. Since we see changes in OCA2 in each case, however, there must have been some advantage in knocking out OCA2, at least in that underground environment. The advantage cannot lie in increased UV absorption, since there's no sunlight in those caves.[4]

There are hints that knocking out OCA2, or at least reducing its activity, may be advantageous (probably in some way unconnected with vitamin D) in humans who can get away with it. We see a pattern that suggests that having one inactive copy of OCA2 is somehow favored even in some quite sunny regions. In southern Africa, a knocked-out version of OCA2 is fairly common: The gene frequency is over 1 percent.[5] Individuals with two copies are albinos and have problems such as skin cancer and blindness as well as social rejection and persecution. Yet it's the most common genetic disease in southern Africa, with the great majority of cases caused by the same mutation. There's a similar story among Amerindians in the American Southwest: A form of OCA2 albinism is common among the Navajo and other neighboring tribes, with gene frequencies as high as 4.5 percent.[6] The same pattern appears in southern Mexico, eastern Panama, and southern Brazil. All of which suggests that heterozygotes—that is, those carrying one copy of the broken version of OCA2—may have some advantage.

Something else that makes us wonder whether vitamin D was the key factor behind recent changes in skin color is the fact, mentioned before, that the genetic changes underlying light skin in Europe and East Asia are almost entirely different. If a reduced-function version of a gene involved in melanin synthesis was strongly favored in Europe, why wouldn't a similar reduced-function version of that same gene arise and spread in China? Mutations that reduce function are quite common. In addition, selection on genes affecting skin color, eye color, and hair color somehow created lots of variety in Europeans: redheads and blondes, blue eyes and green eyes. Nowhere else in the world is that sort of variety common. In most parts of the world, even in temperate regions, everyone has dark eyes and dark hair. To us these facts suggest that there was something fundamentally different in the selective forces affecting skin color in Europe and East Asia. If those forces were different, at least one of them was probably selecting for something other than vitamin D.

## DEM BONES

The skeletal record clearly supports the idea that there has been rapid evolutionary change in humans over the past 10,000 years. The human skeleton has become more gracile—more lightly built—though more so in some populations than others. Our jaws have shrunk, our long bones have become lighter, and brow ridges have disappeared in most populations (with the notable exception of Australian aborigines, who have also changed, but not as much; they still have brow ridges, and their skulls are about twice as thick as those of other peoples.)[7] Skull volume has decreased, apparently in all populations: In Europeans, volume is down about 10 percent from the high point about 20,000 years ago. These changes were spread out over time, of course.

For example, if you look at Bronze Age skeletons from Europe (around 3,000 years ago), you find that some people still had brow ridges like those of Australian Aborigines. Hardly any Europeans have brow ridges today.

Genome-selection surveys may have found some of the alleles affecting these processes. One group of researchers discussed two genes involved in bone growth that showed signs of selection in Europeans, another in the same gene family that showed selection among East Asians, and a fourth that showed signs of selection in both populations.[8] Even though we see similar skeletal changes in many populations, the genetic underpinnings are generally different, much like the pattern underlying skin color.

Some changes can be seen even over the past 1,000 years. English researchers recently compared skulls from people who died in the Black Death (≈650 years ago), from the crew of the *Mary Rose*, a ship that sank in Tudor times (≈450 years ago), and from our contemporaries. The shape of the skull changed noticeably over that brief period—which is particularly interesting because we know there has been no massive population replacement in England over the past 700 years. The height of the cranial vault of our contemporaries was about 15 percent larger than that of the earlier populations, and the part of the skull containing the frontal lobes was thus larger.[9]

## CHEATERS PROSPER

Usually a new version of a gene increases in frequency because it aids the bearer in some way—although it may not aid the species as a whole. Some alleles take this a step farther and succeed by helping themselves, not the bearer. They're called "driving genes."

Everyone has two copies of all chromosomes other than the sex chromosomes, and so everyone has two copies of each gene on those autosomal chromosomes. In the process of meiosis (the process that forms germ cells), a diploid germ cell replicates its DNA and then divides twice, forming four haploid cells each having a complete set of autosomal chromosomes and a single sex chromosome. In sperm production, all four haploid cells become gametes, while in females only one of the haploid cells becomes an egg.

Generally, each of the two copies of each gene has an equal chance of showing up in a gamete. The system is designed (in an evolutionary sense) to give alleles a fair shake. But sometimes, mutation creates a new allele that has a better chance of getting into a gamete—greater than 50 percent. Think of it as a line-cutter. Along the same line, a mutation might increase a gamete's chance of success—say, by making sperm swim faster—and this might be the case for SPAG6 (for sperm associated antigen 6), a gene involved in sperm motility that has apparently undergone a sweep in Europeans over the past few thousand years.[10]

Driving versions of genes must have come into existence more often as human population increased, just as lightning strikes more Texans than Kansans. In a small population, a driving allele would occasionally come into existence and rapidly go to fixation, but the population might spend most of the time in between such sweeps. The much larger populations associated with behavioral modernity, in particular with agriculture, should have generated driving alleles at a rate perhaps two orders of magnitude higher than the populations of the Old Stone Age. Those new driving alleles would not have taken all that long to spread, since they would have grown exponentially in a well-mixed population. Therefore, modern humans should have an unusually large number of driving genes, either recently fixed or on their way to fixation. By the same argument, any species whose numbers have recently soared—for example, after domestication—is likely to also have an unusually large number of driving alleles.

Recent work may have identified such driving alleles. One study[11] found evidence of selective sweeps in a number of centromeric regions—the centromere being a central region that holds together the two halves of the chromosome and that plays a key role in meiosis and in ordinary cell division (mitosis). Since these regions have relatively few active genes, the centromeres themselves may be the target of selection. There is reason to believe that centromere mutations can affect the way in which alleles end up in the egg or in polar bodies, which are dead-end by-products of the egg's division. Any allele with an increased chance of ending up in the egg, instead of a polar body, would have an advantage, possibly a large one. The researchers found evidence of sweeps in eight out of seventeen chromosomes for which data were available. Those sweeps were regional, mainly in the European and Asian samples, which suggests that they came after the expansion out of Africa.

In the long run, a large population would develop more defenses against driving genes as well as more driving genes. But in the short run, just after a dramatic population expansion, driving genes might be both unusually numerous and unusually troublesome, since selective pressures favoring defenses and modifiers only come into existence *after* the driving alleles become common. This isn't just a theoretical concern—it might have something to do with the mysteriously high miscarriage rate in humans. The fraction of conceptions that lead to a healthy baby may be as low as 25 percent, far lower than in most other mammals, and it seems that most of the miscarriages are caused by chromosomal anomalies. An unusually large number of driving genes may play some causal role in this. Too many alleles trying to shoulder their way into the egg at the same time could lead to trouble, just like Stooges all trying to get through the door at the same time.

## CHANGING MINDS

The most interesting kind of genetic changes are those that affect human personality and cognition, and the evidence is good that such changes have indeed occurred.

A number of the new, rapidly spreading alleles found in the recent selection surveys have to do with the central nervous system. There are new versions of neurotransmitter receptors and transporters—neurotransmitters being molecules that relay and influence signals between nerve cells. Several of the new alleles have effects on serotonin, a neurotransmitter involved in the regulation of mood and emotion. Many recreational and therapeutic drugs (particularly antidepressants) modulate serotonin metabolism. And there are new versions of genes that play a role in brain development: genes that affect axon growth, synapse formation, formation of the layers of the cerebral cortex, and overall brain growth. Again, most of these new variants are regional: Human evolution is madly galloping off in all directions.

We see new versions of several genes in factors having to do with muscle fibers and brain function. *Dystrophin* is a protein (coded by the longest of all known human genes) that has an important structural role in muscle fibers and the brain; the *dystrophin complex* is a set of proteins that are physically associated with dystrophin. Major defects in the dystrophin gene itself cause Duchenne muscular dystrophy, which has very severe effects, while lesser defects cause Becker's muscular dystrophy, which is milder. These are among the most common genetic diseases, apparently because the extremely large and structurally complex dystrophin gene has so many ways of going wrong. Dystrophin's dual role has medical consequences, in that boys with Duchenne muscular dystrophy suffer reduced IQ as well as muscular weakness.

The dystrophin-associated sweeping alleles that we see in the selection surveys (which do *not* cause disease) raise the interesting possibility of direct trade-offs between muscle and brain function in the recent past. We have reason to think that humans circa 100,000 BC had stronger muscles than today—and so changes in the dystrophin complex may have sacrificed muscle strength for higher intelligence.

Another very intriguing pattern involves new versions of genes that affect the inner ear.[12] We wonder if this is a consequence of recent increases in language complexity sufficiently recent that our ears (and presumably our brains, throats, and tongues) are still adapting to those changes. Or, since some of the sweeping genes involving the inner ear are regional and recent, could some populations be adapting to characteristics of particular languages or language families? It seems that all humans can learn any human language, but we don't know whether everyone is inherently just as good as everyone else at learning every language, communicating in every language, or eavesdropping in every language.

More generally, these sweeping neurological genes could be responses to the new challenges posed by agriculture itself and the dense hierarchical societies it made possible. In the following sections, we discuss those challenges and likely adaptive responses to them.

## THE MALTHUSIAN TRAP

In *An Essay on the Principle of Population*, Thomas Malthus in 1798 observed that population tends to outrun food supply, since population increases geometrically while food supply increases arithmetically. He wrote:

> The power of population is so superior to the power of the earth to produce subsistence for man, that premature death must in some shape or other visit the human race. The vices of mankind are active and able ministers of depopulation. They are the precursors in the great army of destruction, and often finish the dreadful work themselves. But should they fail in this war of extermination, sickly seasons, epidemics, pestilence, and plague advance in terrific array, and sweep off their thousands and tens of thousands. Should success be still incomplete, gigantic inevitable famine stalks in the rear, and with one mighty blow levels the population with the food of the world.

Imagine that a population of farmers is doing well: They have plenty to eat. It's easy for them to raise more than two children per family—they do so, and the population increases. It continues to increase as long as conditions remain the same. More people need more food, but then there are more workers producing food. As long as per capita production stays the same, the standard of living does not change, even as population increases. However, eventually this expanding population runs out of land, and farmers in the next generation have to farm smaller plots. They may be able to keep per capita production the same by working harder, but in the next generation plots become even smaller. If the methods of food production remain the same, eventually per capita production must decrease as population increases and per capita resources decrease. That decrease will continue until the average farmer produces just enough food to raise two children, at which point population growth stops.

Suppose that farming methods improve, so that productivity per acre goes up by a factor of ten. The population begins to grow—let's say fairly slowly, with each family managing to raise 2.5 children (on average) to adulthood. The population is growing 25 percent per generation. In ten generations—about 250 years—the population has caught up with those improved methods. Living standards are low again, and population growth stops. But 2.5 children per family is by no means an especially high rate of population growth: In colonial America, the average family raised more than 7 children to adulthood. At that rate, population growth could catch up with a tenfold increase in productivity in just two generations.

The point is that even moderate rates of population growth can rapidly catch up with all plausible improvements in food production. Thus, populations should spend most of the time near a Malthusian limit, and there should be no lasting improvement in the standard of living. Malthus himself pointed out that factors other than food shortages can also limit population. Any negative factor that intensifies as population density increases can be the limiting factor—starvation and malnutrition are not the only possibilities. The key is which negative factor shows up at the lowest population density. We believe that the nature of the key limiting factor—which is not necessarily the same in all human populations—can have important effects on human evolution, including the recent changes we have been discussing.

One might imagine that some form of birth control could also effectively limit population, but of course that only works if everyone adopts it. Even smallish groups that do not limit their fertility will rapidly displace (in a few centuries at most) those that do, which brings us back to where we started—a trap where population growth and limitations on population growth keep pace, causing us to remain near the Malthusian limit without achieving lasting improvements in standard of living. In the future, under a disciplined world state, the imposition of birth control could conceivably invalidate the principle, as it could lower the rate of population growth and enable a higher standard of living to take root, but birth control has certainly never worked that way in the past.

War (defined broadly to include all kinds of interpersonal violence) might limit population before starvation occurred if it increased strongly as human density increased. If humans had been unable to form large, well-organized societies, war might have saved us from penury: In fact, war probably has been an important limiting factor in many species other than our own and was probably important for early humans. But humans *can* cooperate, particularly if there is something worth stealing. In a population with a storable surplus, state formation eventually limited local violence—and peace led to the poorhouse.

Infectious disease is the most serious rival to famine as a population limiter. Certainly the two can work together and often have, since malnutrition can lead to reduced disease resistance, while infectious agents can reduce work output—and thus food production. Furthermore, an infectious disease made worse by population density, or one that killed even well-fed people, could, in principle, be the key population-limiting factor in a society. In such a situation, humans would generally have plenty of food, but other trade-offs would be present. There would, for example, be weaker selection for metabolic efficiency than in a classic famine-driven Malthusian trap. Depending on how most of the people made a living, women might become self-supporting and have a reduced need for paternal investment. There would be strong selection for resistance to the organism or organisms responsible for that strong disease pressure.

Each of those three horsemen—war, pestilence, and famine—has dominated in different populations and time periods.

Primitive warfare was apparently the dominant mechanism limiting population among most foragers before the development of agriculture in the Neolithic period. Infectious disease must have mattered in hunting-gathering societies, but its impact was mitigated by foragers' low population density. Strong climatic swings, such as major droughts or cold waves, must have sometimes rapidly reduced the land's carrying capacity and caused famine—particularly during the climatically unstable glacial periods. But, judging from the abundant evidence of homicide and cannibalism in the archaeological record, our guess is that local violence had a stronger effect. In this sort of system, people were egalitarian, and it shows in the genes: The fraction of men fathering the next generation seems to have been markedly higher than in agricultural societies. Infectious disease, in particular falciparum malaria, may have been the limiting factor in tropical Africa. From what we know, it seems that until very recently population densities stayed lower in Africa than in Europe, the Middle East, or East Asia. The female-dominated farming system seen in much of Bantu Africa, in which women were largely self-supporting, indicates that producing food was fundamentally easier there than in most of Eurasia. In much of Eurasia, hard work from two parents barely allowed break-even reproduction. Disease may have limited the complexity of African state systems—but of course there were many other factors, ranging from Africa's relative isolation from rest of the Old World to elephants attacking the fields of pioneers.[13]

In many parts of the Old World, particularly among farmers living under strong states, famine and malnutrition were the main factors limiting population. With internal peace, population rapidly bumped up against carrying capacity. In those societies, people living on the bottom rungs of society were regularly short on food, so much so that they often couldn't raise enough children to take their place. However, elites must have had above-replacement fertility, and their less successful offspring would have replaced the missing farmers. Gregory Clark, in *A Farewell to Alms*, shows that in medieval England the richest members of society had approximately twice the number of surviving offspring as the poorest.[14] The bottom of society did not reproduce itself, with the result that, after a millennium or so, nearly everyone was descended from the wealthy classes. There is reason to think that this happened in many places (eastern Asia and much of western Europe, for example), but wealth was not acquired in the same way everywhere, so selection favored different traits in different societies.

## UNDER THE YOKE

As Rousseau wrote, "Man was born free, and he is everywhere in chains."[15]

In the days before agriculture, governments didn't really exist. Most of the hunter-gatherers were egalitarian anarchists: They didn't have chiefs or bosses, and they didn't have much use for anyone who tried to be boss. Bushmen today still laugh at wannabe "big men." Perhaps we could learn from them.

But farmers do have chiefs: It goes with the territory. Grain farmers store food, and so they have something valuable to steal, which wasn't the case among hunter-gatherers. Elites, defined as those who live off the productive work of others, came into existence in farming societies because they could. Interestingly, some peoples seem to have curbed the growth of elites just by growing root crops such as yams that rot quickly unless left in the ground, and thus are hard to steal.[16] Another point is that the strongest early states often had natural barriers that made it difficult for "citizens" to escape the tax collectors. Egypt, with a strip of very fertile land embedded in uninhabitable desert, is a prime example.[17]

Of course, once your neighbors form states, there's pressure on your group to do the same, both for self-defense and for the benefit of those locals who will form the new elite. Today, practically everyone lives under some kind of government.

Once elites became possible, elite reproductive advantage kicked in. This is the most basic kind of class struggle—the struggle for existence—but it has seldom been noticed by historians, or for that matter by the participants. It could take various forms. In some cases, tremendous advantage accrued to a single male lineage—it's good to be the king! Researchers have found a surprisingly common form of Y chromosome in 8 percent of Ireland's male population. That Y chromosome is also fairly common in parts of Scotland that are known to have had close ties with Ireland, and among the Irish diaspora. Worldwide, 2 million to 3 million men carry this chromosome, and it appears to be the marker of direct male descent from Niall of the Nine Hostages, a high king of Ireland around AD 400. For some 1,200 years (until 1609), his descendants held power in northern Ireland.[18]

The most spectacular example is Genghis Khan, otherwise known as the Scourge of God, the Master of Thrones and Crowns, the Perfect Warrior, and Lord of All Men. About 800 years ago, Genghis and his descendants conquered everything from Peking to Damascus. Genghis knew how to have a good time. Here's his definition of supreme joy: "to cut my enemies to pieces, drive them before me, seize their possessions, witness the tears of those dear to them, and embrace their wives and daughters!"[19] It appears that the last part of that list especially appealed to him. He and his sons and his son's sons—the Golden Family—ruled over much of Asia for several hundred years, tending to the harem throughout. In so doing, they made the greatest of all genetic impacts. Today some 16 million men in central Asia are his direct male descendants, as shown by their possession of a distinctive Y chromosome. It just shows that one man can make a difference.

The elite's reproductive advantage was usually less concentrated. For example, we often see cases in which a relatively small band of conquerors takes over a society and becomes the new elite. If that ruling elite has a strong reproductive advantage and doesn't intermarry much with the original population, the average inhabitant may eventually be largely descended from that elite, even without any obvious or deliberate genocide. This may have happened when the Anglo-Saxons conquered England; although they were greatly outnumbered by the existing Celtic speakers, they account for a large fraction of the modern English gene pool. There is evidence of an apartheid-like social structure in early Anglo-Saxon England that would have furthered this trend.[20]

If it was possible for individuals to move into and out of the elite, which was often the case, traits that increased the probability of entry and continuing membership would have been favored by natural selection. This could happen in any class that had above-replacement fertility, not just in a ruling class. As long as there was significant gene flow, traits favored in that class would tend to increase in the population as a whole, not just in the high-fertility groups.

But if a high-fertility subpopulation was reproductively isolated (or nearly so) for long enough, selective pressures specific to that social niche might cause them to evolve in an unusual direction and become significantly different from the surrounding population. We think this happened among the Ashkenazi Jews, as we discuss at length in Chapter 7. Suffice it to say, for now, that the kind of natural selection that occurred among the Ashkenazim was possible because of the persistence over centuries of strong prohibitions against intermarriage and an odd social niche in which certain traits conferred high fertility. It's a very unusual case, since few populations appear to have experienced the long-lasting reproductive isolation and unusual job mix required to get those results. There are all sorts of ways in which that process could have been interrupted; it's being interrupted now, for example, through high rates of intermarriage between Jews and non-Jews and by changes in fertility patterns.

We've said that the top dogs usually had higher-than-average fertility, which is true, but there have been important exceptions. Remember that rulers, then as now, made mistakes, had bad luck, and in fact often had no idea what they were doing. Sometimes ruling elites lost wars and were replaced by outsiders, as in the Norman Conquest. Sometimes they got a little too enthusiastic about slaughtering each other, as in the Wars of the Roses. And often ruling elites just made bad choices—bad in terms of reproductive fitness, that is. The most common mistake must have been living in cities, which have almost always been population sinks, mostly because of infectious disease. By "population sink," we mean that city dwellers couldn't manage to raise enough children to break even: Cities in the past, before modern medicine and civil engineering, could only maintain their population with a continuing flow of immigrants from the surrounding countryside.

Wealth could make up for the risks that cities presented, if the disease risk wasn't too bad—immunity to famine is an automatic perk of the ruling class, and it's worth quite a lot. But if disease risks were severe, even complete immunity from famine might not be enough, and the ruling elite would gradually disappear—it may not have been obvious, but it sometimes happened.

The disease risks of cities must have gotten worse with time as new pathogens adapted to humans and as civilizations separated by geographic barriers made contact and exchanged pathogens (as happened to the Hittites). We know, for example, that falciparum malaria did not always exist in Italy, but arrived and spread gradually up the peninsula during classical times.[21] Smallpox was also a latecomer to Italy, and it's possible that the addition of those two mighty diseases turned Rome into a population sink for the empire's elites.

Sometimes evolutionarily bad choices on the part of a ruling class are obvious. In classical times, there was a plant called silphium that grew in a narrow coastal strip of Cyrenaica, modern-day Libya. Its resin was used as a contraceptive and abortifacient. The resin appears to have been very effective, preventing pregnancy with a once-a-month pea-sized dose. Silphium eventually became too popular for its own good. Never domesticated, it was overharvested as demand grew. As it became scarcer, the price rose until it was worth its weight in silver, which drove further overharvesting and eventually led to one of the first human-caused extinctions in recorded history. However, during the centuries in which it was routinely used by the Greco-Roman upper classes, it must have noticeably depressed fertility, unless they were throwing money out the window.

Eventually, in some populations, elites turned into governments with a local monopoly of force. You would think that the resulting law and order would have been good for the peasants. They were safer, since they were no longer allowed to raid and be raided by their neighbors. This was a major change, since pre-state warfare often killed a larger fraction of the population than major modern wars do. Peasants still experienced war

Man with the Hoe

*The J. Paul Getty Museum, Los Angeles*
*Jean-François Millet*
*About 1860–1862*
*Black chalk and white chalk heightening on buff paper*
*28.1 × 34.9 cm (11 1/16 × 13 3/4 in.)*

with external foes, but the percentage killed by violence seems to have decreased. However, since births and deaths still balanced, every decrease in death by violence was counterbalanced by an increase in deaths caused by infectious disease (which hit everybody, including elites) and starvation (peasants only). Government, especially good government, eventually led to decreased standards of living, at least in terms of calories.

## WE FOUGHT THE LAW (AND THE LAW WON)

If your ancestors were farmers for a long time, you're descended from people who decided it was better to live on their knees than to die on their feet.

Farming led to elites, and there was no avoiding their power. Foragers could walk away from trouble, but farms were too valuable (too important to the farmers' fitness) to abandon. So farmers had to submit to authority: The old-style, independent-minded personalities that had worked well among egalitarian hunter-gatherers ("A Man's a Man for a' That") were obsolete.[22] Even when some group had a chance to refound society on a more egalitarian basis, as in the case of the medieval Iceland republic, elites tended to reappear.[23]

Aggressive, combative people may also have experienced lowered fitness once ruling elites began to appear. With strong states, the personal payoff for aggression may have become smaller, while law and order made combativeness for self-defense less necessary. Sheer crowding must also have disfavored some personality traits that had worked in the past. Intuitively, it seems that a high level of aggressiveness would be less favored when encounters with strangers were frequent. Fight too often and you're sure to lose. Moreover, although the winner of a deadly struggle between two peasants might conceivably gain something, his owners, the elites who taxed both of those peasants, would not, any more than a farmer benefits when one bull kills another.

Farmers don't benefit from competition between their domesticated animals or plants. In fact, reduced competition between individual members of domesticated species is the secret of some big gains in farm productivity, such as the dwarf strains of wheat and rice that made up the "Green Revolution." Since the elites were in a very real sense raising peasants, just as peasants raised cows, there must have been a tendency for them to cull individuals who were more aggressive than average, which over time would have changed the frequencies of those alleles that induced such aggressiveness. This would have been particularly likely in strong, long-lived states, because situations in which rebels often won might well have favored aggressive personalities. This meant some people were taming others, but with reasonable amounts of gene flow between classes, populations as a whole should have become tamer.

We know of a gene that may play a part in this story: the 7R (for 7-repeat) allele of the DRD4 (dopamine receptor D4) gene. It is associated with Attention-Deficit/Hyperactivity Disorder (ADHD), a behavioral syndrome best characterized by actions that annoy elementary school teachers: restless-impulsive behavior, inattention, distractibility, and the like.

The polymorphism is found at varying but significant levels in many parts of the world, but is almost totally absent from East Asia. Interestingly, alleles *derived* from the 7R allele are fairly common in China, even though the 7R alleles themselves are extremely rare there. It is possible that individuals bearing these alleles were selected against because of cultural patterns in China. The Japanese say that the nail that sticks out is hammered down, but in China it may have been pulled out and thrown away.

Selection for submission to authority sounds unnervingly like domestication. In fact, there are parallels between the process of domestication in animals and the changes that have occurred in humans during the Holocene period. In both humans and domesticated animals, we see a reduction in brain size, broader skulls, changes in hair color or coat color, and smaller teeth. As Dmitri Belyaev's experiment with foxes shows, some of the changes that are characteristic of domesticated animals may be side effects of selection for tameness. As for humans, we know of a number of recent changes in genes involving serotonin metabolism in Europeans that may well influence personality, but we don't know what effect those changes have had—since we don't yet know whether they increase or decrease serotonin levels. Floppy ears are not seen in any human population (as far as we know), but then, changes in the external ear might interfere with recognition of speech sounds. Since speech is of great importance to fitness in humans, it may be that the negative effects of floppy ears have kept them from arising.

Some of these favored changes could be viewed as examples of neoteny—retention of childlike characteristics. Children routinely submit to their parents—at least in comparison to teenagers—and it's possible that natural selection modified mechanisms active in children in ways that resulted in tamer human adults, just as the behaviors of adult dogs often seem relatively juvenile in comparison with adult wolf behavior.

If the strong governments made possible by agriculture essentially "tamed" people, one might expect members of groups with shallow or nonexistent agricultural experience to be less submissive, on average, than members of longtime agricultural cultures. One possible indicator of tameness is the ease with which people can be enslaved, and our reading of history suggests that some peoples with little or no evolutionary exposure to agriculture "would not endure the yoke," as was said of Indians captured by the Puritans in the Pequot War of 1636. In the same vein, the typical Bushman, a classic hunter-gatherer, has been described as "the anarchist of South Africa."

## BOURGEOIS VIRTUES

Agriculture itself, and the particular form it took in state societies, must have selected for personalities that can only be called bourgeois, characterized by the traits that make a man successful rather than interesting. One such trait was the ability to defer gratification for long periods of time. This was a practical requirement for farmers, since they had to save a portion of their crop for seed and some of their domesticated animals for breeding stock.

This wasn't easy. Food was often shortest just before sowing, and those early farmers had to abstain from eating the seed grain when they and their families were hungriest. This is something that classic hunter-gatherers just didn't do: There was no way for them to store food effectively, so they either consumed it on the spot or shared it with others. Foragers had no tradition of self-denial and no inclination to deny themselves. They weren't very good at self-denial back in the early Neolithic period, and they aren't very good at it even today: Efforts to teach Bushmen to become herders frequently fail when they eat all their goats. People can learn new traditions, but genetic differences must make this kind of self-denial easier for some people than it is for others. It takes a certain type of personality—with traits including patience, self-control, and the ability to look to long-term benefits instead of short-term satisfaction—and natural selection must have gradually made such personalities more common among peoples that farmed for a long time.

Agriculture also led to the birth of property. Among hunter-gatherers, there hadn't really been any. Although tribes sometimes claimed hunting grounds, there was no individual ownership of land. The mobile way of life that hunter-gatherers pursued kept them from accumulating much property other than some personal tools and weapons. Farmers, being sedentary, could accumulate domesticated animals, land, and other forms of property. This became more practical and more important as states appeared and limited local violence. Law and order allowed for population gains that increased the scarcity and value of land. In some cases, governments made property safer and more secure.

Farmers could thus accumulate resources that increased their fitness and that of their descendants—if they decided to do so, and if the state didn't take too much. But those choices didn't come easily. Hunter-gatherers routinely *shared* resources, partly in order to cement relations with other members of the tribe, partly because there often wasn't anything else to do with those resources. Try eating a whole giraffe before the meat goes bad. Even with the wife and kids helping, it can't be done. The effective cost of sharing that meat is zero. Foragers aren't selfish.

Farmers, in contrast, have to be selfish. At minimum, they can't afford to give away seed grain or breeding stock—not if they want to stay farmers. More than that, farmers could gain increased fitness by being miserly, at least in comparison to foragers.

And once there was property, laziness must have decreased. There were many ways in which hard work could produce enduring assets that could increase an individual's fitness or that of his children and relatives. Farmers could save to buy more land or livestock. They could build long-lasting improvements like buildings or irrigation works. This was not really possible for hunter-gatherers—there was no way for them to accumulate wealth. If they had full stomachs and their tools and weapons were in good shape, hunter-gatherers didn't work. They hung out: They talked, gossiped, and sang. They were lazy, and they should have been: Being lazy made biological sense. They could usually obtain enough food fairly easily, since constant local violence kept human numbers below the land's carrying capacity. When law and order let human density increase, farmers eventually had to work harder and harder just to survive. Here again, selection must have favored those odd people who *like* to work, even when there was enough to eat.

Ultimately, this meant that both sexes had to work hard. In fact, for most people, that became the only way to produce enough to feed and raise a family. That pattern is not universal. In situations where resources are abundant, men sometimes do little work. Men working hard to feed their families—"high paternal investment," we call it—is common among contemporary hunter-gatherers and may well have been a standard feature of the ancestors of all modern humans. Women bring in most of the calories in such societies (from plant foods), at least in warm climates, but the meat contributed by male hunters is a vital source of protein and other essential nutrients.

However, the grueling labor required of peasants in well-governed states—inevitable when local violence no longer keeps the population well below carrying capacity—takes this to another level entirely. Since a hardworking husband was essential, in some cultures the practice of dowry arose, so that a farmer with assets could buy his daughter a productive husband—making dowries yet another way of using property to increase fitness.

Given a stable government and reasonably low taxes, self-denying individuals could make wealth generate more wealth. In many early civilizations, the real interest rate was around 10 percent per year, a rate high enough that anyone who managed to put some money aside as savings would, after a few decades, be able to kick back and relax a bit. At that point men were too old to truly enjoy that income, but it could still increase their children's fitness, and so selection may have favored that kind of behavior. As "time preference" declined (that is, as propensity toward delay of gratification increased), interest rates eventually fell as more and more people saved rather than spending any spare money.

Agriculture in Malthusian conditions must have also favored individuals who were metabolically efficient and could produce the maximum work for a given amount of food. Among hunter-gatherers the selective pressures were different—in their way of life, bursts of strength in war or hunting were relatively more important. We see differences in the gene alpha-actinin-3 (ACTN3) that may reflect this. The gene has two forms—one that produces a protein that is active in fast-twitch muscles, and one that produces no protein at all. The intact version of the gene increases muscle power and is noticeably more common among world-class sprinters than in the rest of the population; the other version of the gene increases aerobic efficiency and endurance. Gene-engineered mice with the endurance version of ACTN3 can run 33 percent farther than standard lab mice before exhaustion sets in.[24] Both forms of ACTN3 are found in all populations, but the endurance form appears to have become more common since the advent of agriculture in Europe. We suspect that it made peasant farmers more productive.

At first, all these pro-agricultural behaviors must have run against the grain: It's unlikely that humans were comfortable doing things that had never made sense in the past. But over time, alleles that induced this kind of ant-like behavior must have increased in frequency, until eventually, after millennia, selfish, hardworking, self-denying people were far more common than they had been among hunter-gatherers. Acting like ants rather than grasshoppers didn't improve the average standard of living over the long haul, since the world was Malthusian, but farmers who worked harder than average and saved more than their neighbors would have had higher-than-average fitness. Eventually there must have been many people with personality types that hadn't existed at all among our forager ancestors.

## MONEY AND MARKETS

In *Narrow Roads of Gene Land*, theoretical biologist William Hamilton wrote, "It seems to me that there are some aspects of innate intelligence that civilization steadily promotes. Mercantile operations, for example, are an inseparable part of Old World civilizations and need complex models in the minds of their operators, just as military ventures do. The main difference is more emphasis on prudence and less on daring. It is probable that civilization has given steady selection for the intelligence needed for this mercantile kind of preparatory modeling."[25]

Agriculture would have selected for traits that enable people to engage successfully in trade: A farmer able to sell his wheat at a higher price than the other wheat farmers or to make more advantageous trades would have been more successful and better able to support a large family. And so salesmen, businessmen, and financiers were born.

If this theory is correct, we would not expect populations that have never experienced such pressures, or that have only experienced them for limited periods and to a limited degree, to be very successful in such activities today. Groups that became agriculturalists relatively recently, or not at all, are slow to master important new social and technical developments. This is the case for the Amerindians, and it underlies a current wave of discontent with liberal economic policies in South America.

Along the same line, the well-known middleman minorities, such as Armenians, Jews, Lebanese, Parsees, Indians in East Africa, and Chinese in Southeast Asia, are all descended from long-established agricultural populations. Evidently, a long history of getting and spending does not lay waste your power.

## MODERN TIMES

The kind of gradual, directional biological change we've outlined should have generated historical trends. These trends wouldn't necessarily have been monotonic, but over time, to the extent that the underlying biological changes favored certain kinds of societies and organizations, those kinds of societies and organizations would have become more common. Ultimately, some populations may have changed enough to allow some social patterns to prevail that couldn't have worked at all in long-ago societies. More exactly, those patterns could only exist in populations that had been reshaped over millennia by the selective pressures associated with hierarchical agricultural societies.

The fact that there is undoubtedly a lot of overlap between the psychologies of different populations does not mean that the same social patterns and the same kinds of organizations are always possible. The *distribution* of personality traits also matters.

For example, a high-trust society can largely avoid some costs—in modern terms, they could leave their doors unlocked and wouldn't worry much about corruption. All else equal, they'd be more effective in war than societies with less trust among their own members, since they wouldn't suspect betrayal at every setback. A different society, one in which 20 percent of the population practiced cheater strategies, would have to spend a lot of resources on punishment and prevention. Certain activities that required a high degree of mutual trust might be effectively impossible.

So a population in which no personality was truly alien (one in which everyone resembled some character in Shakespeare, for example) might generate a qualitatively different civilization because the *mix* of personality types differed from ours. A society that had many more Hamlets than ours might never accomplish anything at all.

We have two ways of looking for social patterns favored by recent natural selection. The most obvious is to look for patterns whose frequency changed over time. In the strongest examples, such patterns would be rare or unknown until some point in history, possibly quite recently. This can be difficult to do, though, because in many cases we simply do not have much information about ancient civilizations. For example, we doubt that the Indus civilization had a bicameral legislature, an independent judiciary, and a written constitution—but since we can't read their script, how can anyone know? The other way is to exchange time for space: to look at contemporary peoples that have never lived as peasants, or have done so for a considerably shorter period of time than Europeans, East Asians, and Middle Easterners, then check to see what social patterns and institutions (if any) do not flourish in those populations. Although this method can be more controversial than looking at the earliest civilizations, it does have one advantage in that recent history is at least well documented.

The relative ease with which old agricultural civilizations (many of them, anyhow) have managed to adopt complex new technologies and forms of social organization, compared to populations that have had less experience with agriculture and dense hierarchical societies, suggests that gradual biological changes in cognition and personality played a key role in the birth of the industrial and scientific revolutions.

Jared Diamond observed in *Guns, Germs, and Steel* that "the nations rising to new power are still ones that were incorporated thousands of years ago into the old centers of dominance based on food production, or that have been repopulated by people from those centers. . . . Prospects for world dominance of sub-Saharan Africans, Aboriginal Australians, and Native Americans remain dim. The hand of history's course at 8000 BC lies heavily on us."[26] But what kind of experience are we talking about? Diamond asserts that such differences were entirely cultural, that is to say, learned—but if this were so, populations that missed out on these experiences could in principle catch up rapidly. After all, culture is learned anew every generation, so presumably new technologies and new forms of social organization that had proved successful in other countries could be adopted over two or three generations, just as most farmers in many countries have become city dwellers over a few generations. Yet economists have shown that the age of the transition to agriculture appears to have a strong influence on a country's economic development in recent decades, even after controlling for many other factors.[27] It's hard to see how this could be due to cultural effects. Even if a nation could learn from experiences their ancestors had back in the Bronze Age and benefit from them (which sounds unlikely), why can't everyone else learn those same lessons? Why would those experiences confer a relative advantage? On the other hand, genetic changes that accommodated people to a dense hierarchical society could easily have developed over those millennia, and genetic information can't easily be transferred—yet.

If the root causes of these differences are biological changes affecting cognitive and personality traits, changes that are the product of natural selection acting over millennia, conventional solutions to the problem of slow modernization among peoples with shallow experience of farming are highly problematic. And yet, methods based on an understanding of underlying biological causes might be very effective. It's entirely possible that such methods would turn those prospects of world dominance around, which would certainly liven things up.

The new social patterns we find most intriguing are those that have led to greatly increased rates of innovation in the past few centuries—usually called the scientific and industrial revolution. Some argue that gradual genetic changes could not be responsible for such rapid social changes. We don't call them revolutions without reason. We believe, however, that these arguments are mistaken. Consider an example in which an allele affecting behavior had a frequency of 20 percent and a 6 percent selective advantage in a European population in 1500—we know that there are many sweeping alleles with a selective advantage in that range. Over the next 300 years, the frequency of that allele would have doubled, and going from 20 percent to 40 percent could be a significant change, enough to give European society in 1800 some new capability or tendency.

Such a favored allele takes about a millennium to increase by a factor of ten. When its frequency is 1 in 100,000, 1 in 10,000, or even 1 in 1,000, that allele has no social impact: But when the frequency changes from 1 percent to 10 percent, the allele starts to make itself felt. When it increases from 10 percent to 50 percent in less than a millennium, its impact could be dramatic.

Modest biological changes might also trigger dramatic social changes by crossing some threshold, just as a tiny increase in temperature can turn ice into water. Such changes (ice into water, water into steam, graphite into diamond) are called *phase transitions*. There may be analogous social transitions. Picture an army in the middle of a battle, one that's not going very well. Soldiers are beginning to run away—first a few, then more. Those who are still fighting suspect that their chance of victory is rapidly decreasing as more and more of their comrades leave, and as the odds grow worse, more and more soldiers flee. This accelerates until the army completely disintegrates, with every soldier trying to save himself. A small change in the battle situation has transformed a highly organized, functional army into a mob. Depending on the mix of personality types in that army, such disintegration could range from unlikely to virtually inevitable, and the difference between those mixes might not be all that large. Cultural factors could influence the probability of that kind of social transition, but so could biological influences on personality.

It is also likely that some significant activities had effective thresholds, such that only individuals with atypical traits could perform them. It's easy to imagine a boulder so heavy that only a few of the strongest men could lift it, but then it's also easy to imagine a puzzle so difficult that only a few people could solve it, or a song with notes so high that only a few people could sing it. In such situations, outliers are important.

Many traits are distributed approximately in a bell curve, or "normal distribution." This is the shape that describes most of us being middling, a few of us being a bit different from average, and a tiny number of us being quite a lot different from average. The average height among men in the United States, for example, is about 5 feet 9 inches, while the standard deviation (the typical difference between two U.S. men chosen at random) is about 3 inches. This means that about two-thirds of men are between 5 feet 6 inches and 6 feet tall, with about a sixth over 6 feet. As we go further from the norm, we find fewer and fewer individuals: About 1 in 50 are over 6 feet 3 inches, while one in 770 are over 6 feet 6 inches. The fraction above a threshold falls off more and more rapidly as we increase the threshold. Now consider another, shorter population—let's say that the average height of men in this population is 5 feet 6 inches, one standard deviation below the average height of American men. There is substantial overlap between the two populations, but the difference in the frequency of Eastwoods becomes very large: Men taller than 6 feet 6 inches will be more than forty times rarer in the shorter group than in the U.S. sample.

The point here is that a modest difference in the mean of some trait can have a tremendous effect on the frequency with which members of a group exceed a high threshold. If some important cultural task can only be accomplished by individuals who are unusually good at solving certain kinds of puzzles, then the course of cultural evolution may change radically with modest changes in the group's average puzzle-solving ability. There are many other factors that might influence such events, but a difference in mean ability due to genetic differences is one of them. And both of these factors—social phase transitions and increases in the frequency of people with specific talents—may have played a part in the birth of modern science.

Science as we know it got its official start in Europe in the sixteenth century with the publication of Copernicus's work *De revolutionibus* in 1543. The closest thing to modern science seen before that would have been the protoscience practiced by the Greek and, later, Arab civilizations—but they're not that close. The productivity and intensity of modern science far outshines earlier efforts. Some of the most important European scientists, such as Isaac Newton, James Clerk Maxwell, and Charles Darwin, made larger intellectual contributions *as individuals* than other entire civilizations did over a period of centuries.

We believe that science requires communication and cooperation between people who are unusually good at (and interested in) puzzle-solving. Science is a social enterprise, and scientists never truly work alone: They always build on the work of others. It was Newton who said, "If I have seen further it is by standing on the shoulders of Giants," and he ought to have known. So the number of such people, and their social connections, is crucial to the progress of science. We also know that modest differences in mean ability can have a big effect on how common such people are.

You see, there can also be phase transitions in *connectivity*. Imagine that the average budding scientist in Europe in 1450 knew a few other people like himself. Those acquaintances knew others, but since such people were rare, the potential scientists of Europe fell into small, isolated groups rather than a single connected community. There was no efficient way for new ideas and discoveries to spread. We are positing that as the frequency of such people increased, there was a sharp transition at a certain critical value. Suddenly all groups connected, and there was a path between any two members. Something similar happens in epidemiology: If the number and density of vulnerable individuals exceeds a certain threshold, the infectious disease is certain to spread to the entire community. Below that threshold, the disease is confined to a small cluster of people and dies out.

Thus, the scientific "revolution" may well have resulted from modest changes in gene frequencies affecting key psychological traits. What traits would have favored the birth of science? Increases in abstract reasoning or numerical abilities might have helped, and it's possible that those traits were favored by selection in complex, hierarchical societies. Generally, though, we think there was no direct selection favoring creativity itself, and that creative individuals are accidental by-products of selection for other traits, traits that really did pay off in everyday life, such as low time preference and the ability to make complex mental models.

Our view is in sharp contrast to those who have argued that creativity conferred fitness benefits. It has been shown that poets are unusually likely to be manic-depressive.[28] Building on this, others have argued that alleles underlying manic-depression should have increased in frequency because of the social rewards received by poets and other creative artists.[29] Of course, few people carrying those alleles had a chance to be poets: Most (in recent millennia) must have been hardscrabble farmers, and it's hard to see how manic-depression could have been an advantage in that situation.

In fact, poets have seldom received large rewards, and their fitness has often been low—particularly among those with manic-depression, as a result of its high suicide risk. More generally, creativity seldom confers large fitness advantages, because good new ideas can be rapidly copied by others. The copiers receive the fitness benefits without paying the associated costs. In fact, it's been obvious for a long time that innovators seldom harvest much of the benefit generated by their innovations. Public policy has aimed at increasing those rewards—for example, through patent systems and public support of scientific research. Such support is limited and fairly recent, however, and over the long run of human history and prehistory, direct selection for creativity seems unlikely.

Technical and social factors must have been important in increasing social connectivity: Better transportation, regular mail services, and the printing press, for example, played essential roles. Although inventions such as the printing press were undoubtedly important, they seem to have been necessary rather than sufficient, since science either does not exist or is appallingly feeble in the majority of the world's populations, even among those that have access to those favorable technological factors. If a region or population produces major advances in knowledge, science there is real and alive, otherwise not. By that standard, science does not exist in sub-Saharan Africa or in the Islamic world today. As Pervez Hoodbhoy (head of the physics department in Islamabad) has written, "No major invention or discovery has emerged from the Muslim world for well over seven centuries now."[30]

Although we do not as yet fully understand the true causes of the scientific and industrial revolution, we must now consider the possibility that continuing human evolution contributed to that process. It could explain some of the odd historical patterns that we see. For example, if people hadn't yet changed enough, the failure of Hellenistic science to take off may have been inevitable. In addition, such ideas may help explain why some populations with an early start on agriculture and state formation have found it easy to participate in these revolutions, while those with late starts have not. In particular, we think that the story of the Ashkenazi Jews, many of whom have played important parts in the later phases of those two revolutions, was shaped by this kind of evolution—evolution over historical time.

# 5
# GENE FLOW

## GENETIC HISTORY

Geneticists have traditionally traced the flow of genes in order to study the movements and origins of peoples. They've studied particular variants of the Y chromosome in an attempt to determine which ethnic group in Asia is most closely related to the Amerindians.[1] They've tried to determine the extent to which modern Europeans are descended from ancient Europeans of the Upper Paleolithic period who adopted farming, or from Neolithic immigrants from the Middle East.[2] Researchers have used mitochondrial DNA (mtDNA) and Y-chromosome data to determine which groups contributed maternal and paternal ancestry to a mixed population—they have determined, for example, that most Mexican Y chromosomes are of Spanish origin, whereas most Mexican mtDNA is Amerindian.[3] They have tried to analyze other ancient population movements in this way as well, most notably the original human expansion out of Africa.

There are two ways of looking at these types of informative gene variants. In the kind of analysis conducted to determine paternal and maternal ancestral lines, on the one hand, researchers are only interested in these gene variants as markers of past population movement and admixture rather than in the functions of the variants themselves. The assumption is that one Y chromosome functions just like any other and that all mtDNA variants have the same properties: That is to say, they're neutral. But if their properties varied—if the bearers of some Y-chromosome variants had noticeably higher fitness than those bearing other variants—this whole brand of analysis would be thrown into question, particularly when used to look far back into prehistory.

We, on the other hand, are interested in alleles because of their *effects*, precisely because they do make a difference. Generally, we're interested in how population movements and admixture have helped to spread new adaptive variants rather than in how using the variants can help us to track the movements.

Every new mutation, including any rare but important beneficial mutation, starts out as a single copy in one individual. It's local. If it's going to be important, if it's ever going to influence a significant fraction of the human species, it must first spread. Looking at the bigger picture, we can see that the flood of favorable mutations involved in the recent acceleration of human evolution will have major impacts only if they spread widely. Presumably, if our theories are correct, many of them are still in the midst of sweeping through large segments of our world population. This means that the average person today bears many of these new favorable mutations. You are, most likely, significantly different genetically from your ancestors of a few thousand years ago. And although natural selection hardly operates in a way guaranteed to maximize our convenience, it is still the case that many adaptive changes have welcome results. It's hard to argue against something that keeps you alive.

## BREEDING LIKE RABBITS

The settler Thomas Austin released 24 wild rabbits on his Australian farm, called Barwon Park, in 1859, and some other Australian farmers later followed his example. Rabbits are sexually mature at about six months, and they have a 31-day gestation period. Given a favorable environment, rabbits can easily increase their population fourfold in a year. Try to imagine the growth of the rabbit population in Australia: first 24 rabbits, then 100 rabbits after a year, 20,000 in five years, and 25 million after ten years. That's roughly what happened: At the end of ten years, shooting or trapping 2 million a year had no noticeable effect on their population.

At first growth was slow—an increase of 75 in a year doesn't sound that impressive, not in a country the size of the lower 48. But growth speeded up as the rabbit population increased. Another way of putting it is that the percentage of growth per year stayed the same, but that the percentage was multiplied by a larger and larger population as time passed. A process that at first seemed unimpressive left a whole continent literally swarming with rabbits in a single decade. It took only two or three times longer to fill a continent with rabbits than it had to fill a single farm.

A favorable allele, such as the one that confers lactose tolerance, spreads in much the same way, although the process takes thousands of years, largely because human generations are much longer than rabbit generations. But for a sweep to happen rapidly, the population must be "well mixed." This is not always the case, because mixing genes over long distances—over rivers, mountains, deserts, and oceans, or through hostile tribes—is far from automatic. Sometimes it happened, sometimes it didn't. These sweeps were strongly influenced by history—and they influenced history right back.

## HOW A SWEEP BEGINS

Every selective sweep starts out as a change in the DNA of a sperm or egg. Such changes can be caused by chemicals, radiation, or just random jostling of molecules—but what matters to us is that such changes do occur. Mutations favorable enough to initiate a sweep are extremely rare. One set of human DNA has about 3 billion nucleotides, and an average person has about 100 new mutations. Most of those changes are in DNA that apparently does nothing at all—only 2 percent of our DNA does anything (as far as we know)—but on average, two or three of those mutations affect functional DNA. Still, they do not usually make a significant difference, either in a positive or a negative way.

When a mutation does make a significant difference, the effect is almost always negative: Random changes in an incredibly complex piece of machinery are likely to screw things up. Sometimes a change in a single nucleotide can kill or cause serious disability. For example, achondroplasia, the most common kind of dwarfism, is caused by a change in a single nucleotide on chromosome 6—almost always the same exact change. Such negative alleles never become common: Their bearers have fewer children than average, so there will be fewer copies in the next generation. Very rarely, a mutation happens that has a positive effect—a good difference. These rare but supremely important events are the raw material of evolution.

## LIMONE SUL GARDA

In 1980, Italian researchers found that a man from Limone sul Garda (a small lakeside village in northern Italy) had very low levels of HDL ("good" cholesterol) and high levels of triglycerides, yet showed no sign of heart disease. Both of his parents

Limone sul Garda

had lived to advanced ages. Their curiosity whetted, the researchers performed blood tests on all 1,000 inhabitants of Limone and found a total of 43 people with this same unusual blood-lipid profile. The local church had birth records going back centuries, and the researchers were able to determine that all those individuals could trace their ancestry back to the same couple (Giovanni Pomaroli and Rosa Giovaneli), who had married in 1780.[4] This genealogical pattern suggested that these villagers shared a mutation, which turned out to be a change in the protein called ApoA-I (Apolipoprotein A-I), a major component of high-density lipoprotein (HDL). ApoA-I helps to clear cholesterol from arteries, but this variant, ApoA-$I_M$ (M for Milano), apparently does a considerably better job of it. A change in a single nucleotide modified an amino acid in the protein, completely changing its chemical action.

ApoA-$I_M$ is much more effective at scouring out arteries than the standard version of the protein is, and carriers have substantial protection against atherosclerosis. They have a much-reduced risk of heart attacks and strokes, and they often reached an advanced age.[5] Not only that, these effects of the ApoA-$I_M$ mutation have been duplicated in mice, and it protects them against artery plaque as well.[6] Preliminary tests show that intravenously administered synthetic ApoA-$I_M$ actually shrinks preexisting artery plaque in humans: Nothing else we know of does that.

Judging from the records we have, this mutation seems to have increased in number, from 1 copy to 43 in ten generations. Chance and general population growth must have played a role, but let's suppose that freedom from heart attacks and strokes is driving a gradual increase. What would happen if it were given several thousand years to expand?

Let's say that its true advantage—the long-term average—is 7 percent, so that carriers raise 7 percent more children than average. In that case, you'd expect most Europeans to have a copy in 6,000 years or so. This assumes, of course, that Europe will still exist thousands of years from now, that we won't have developed a universal cure for atherosclerosis in the meantime, and that the robots won't have taken over first. We know the future is uncertain—bear with us.

Success in 6,000 years is nothing to hold your breath about, but the point is that mutations with a similar advantage that started in a single village at the dawn of recorded history have had time to become common in just this way. That estimate assumes that genes (and people) are well mixed, but that's clearly not the case in Limone sul Garda. The village is quite isolated: The mountains and the lake hem it in, and there wasn't even a road until the 1930s. Such isolation doesn't make the occurrence of a favorable mutation more or less likely, but the concentration of carriers in one village may have made them easier to notice. However, it certainly interferes with the spread of the gene.

So how did a favorable mutation spread thousands of years ago?

## THE GIRL NEXT DOOR

Few villages today are as geographically isolated as Limone: Most have other villages fairly close to them, and generally there's traffic between them. People make frequent visits to these places close to home. The simplest and oldest mechanism of gene flow—marrying someone from the next village over—therefore still prevails. More often than not, this has meant women leaving their homes to join their husbands' communities, an ancient pattern that we still see in chimpanzees. This village-to-village contact has been a factor in gene flow ever since people settled in villages, and it has been one of the most important ways in which favorable alleles have spread. Given time, neighborhood marriages can carry an allele thousands of miles. A beneficial allele originating in one band or village could, through such intermarriage, gradually spread to neighboring populations, and then to neighbors' neighbors, and so on. Alleles with a big advantage would spread more rapidly than alleles with a small advantage.

In any case, with some simplifying assumptions, it's possible to model the spread of an adaptive allele with a mathematical formula. In that model, the frequency of the favored allele spreads in the form of a wave with a constant speed. The speed depends on the selective advantage and the root-mean-square distance separating the parents' and the child's birthplaces. If we call that marital distance $\sigma$ and the selective advantage of the allele $s$, the speed of advance is approximately $\sigma \times (2s)^{1/2}$ miles per generation.

Hunter-gatherers can be amazingly mobile, and since most recent hunter-gatherers were spread very thinly, there often *were* no girls next door. So hunter-gatherers, especially in sparsely settled areas, had to find mates at a considerable distance. A generation ago, when many Bushmen were still wandering freely, their average marital distance was over 40 miles. This may not have been typical in prehistory. In the days before agriculture, when everybody and his brother was a hunter-gatherer, most lived in choice territories, not in the marginal habitats like the Kalahari Desert where that way of life has persisted. Population density would have been higher in those conditions than among Bushmen today, and people may not have had to search so far for a mate. However, it is clear that agriculture eventually led to crowding. Peasant farmers usually marry people living nearby, not least because there *are* plenty of people living nearby to choose from. In an example discussed by Alan Fix, based on census records from a densely settled part of rural England about 150 years ago, the average marital distance was only 6 or 7 miles.[7]

Consider a new allele that has an advantage of 5 percent. In a well-mixed population it would rise to high frequency in about 8,000 years. Among hunter-gatherers like the Bushmen, it might spread 9 miles per generation, on average, and among farmers, about 1 or 2 miles per generation. Since the preponderance of recent evolution seems to have been driven by the changes associated with agriculture, the 1.5 miles per generation you'd expect in farmers would be a good estimate. So in the 400 generations since the birth of agriculture (at twenty-five years per generation), a gene with a 5 percent advantage would have moved out about 600 miles.

Although this way of spreading genes is simple, universal, and easy to understand, it's not the only way, and it's slow. When you run the numbers, it's hard to see how it can carry alleles as far as they have actually traveled in the time available. There is a similar problem in understanding the spread of oak trees in England. Some 15,000 years ago, back in the Ice Age, oak trees were extinct there, or nearly so—oak trees just don't grow very well under thousands of feet of ice. Possibly a few hung on in protected southern river valleys. Yet today they're all over the island—a fine thing, no doubt, but how did they manage it? Oaks shouldn't spread very fast for the simple reason that the acorn doesn't fall far from the tree.

The answer is that unusual events took place often enough for oak trees to spread faster than would have been possible if all the acorns stayed near their trees. Occasionally a bird carried an acorn far ahead of the main wave of advance and started a new oak forest. An acorn might have floated a long way down a river and sprouted. Maybe some of the first humans who resettled Britain after the ice melted took some acorns with them on a long hunting trip north and dropped a few. The speed of advance is determined more by these rare, long-distance events than by gravity and squirrels.

So although local marriage is a big part of the story, rarer, weirder, more complicated events most likely determined the speed of advance of new favorable alleles in humans.

## BARRIERS TO GENE FLOW

Of course, there were a few major factors that blocked or slowed gene flow over the past few millennia.

The Atlantic and Pacific oceans were important barriers. There was very little contact between the peoples of the Old and New Worlds before Columbus. Australia was much easier to reach from Indonesia or New Guinea than the Americas were. There were definitely visiting sailors from Indonesia fishing for sea cucumbers along the north coast of Australia beginning around 1720. There must have been other early contacts, but the amount of gene flow was not large, judging from what we know of Y-chromosome and mtDNA variants in Australian Aborigines.[8] The north coast did not attract settlers from Indonesia or Southeast Asia, probably because that coast was unsuited to their forms of agriculture. Moreover, new alleles that were adaptive in the context of agriculture may not have had an advantage in Australia: Even if some were introduced, they may not have spread widely.

Deserts mattered. There was plenty of gene flow between North Africa and the other lands surrounding the Mediterranean, particularly after the development of sailing ships, but the Sahara certainly interfered with movement to and from sub-Saharan Africa. The block wasn't absolute. The Sahara was a much friendlier environment in the early Neolithic than it is today. Later, the domestication of the camel favored trans-Saharan trade, while the European and Arab slave trade eventually brought African alleles to other parts of the world.

Still, we know that the amount of gene flow into sub-Saharan Africa was limited in the past, since several local mutations that cause lactose tolerance became common among the cattle-raising peoples in the Sudan and Ethiopia, even though the European version is considerably older. If there had been much gene flow into sub-Saharan Africa back then, the European mutation would most likely have dominated. In fact, if even one person carrying that European allele had successfully settled in the Sudan back in the Bronze Age, he would have had a fair chance of introducing it, as a kind of genetic Johnny Appleseed. The Sahara Desert made such contacts rare, but it may also be that tropical diseases such as malaria and yellow fever interfered, just as they later interfered with European colonization attempts. The local sub-Saharan Africans had enough resistance to those diseases to get by, but outsiders generally did not.

Most mountain ranges affect gene flow but aren't impassable enough to substantially block it. The Himalayas, however, are an exception. Judging from the limited information we have today, it seems that India shares a fair number of favored new alleles with Europe and the Middle East, but has mixed far less with China. Since the Himalayas are the tallest mountains on earth, backed by the Tibetan Plateau, it's possible to believe that they greatly reduced gene flow between India and China, and ultimately between east and west Eurasia.

Other people getting in the way must have been one of the most powerful forces slowing gene flow. Many alleles that helped people adjust to agriculture probably came into existence in the Middle East, since agriculture began there. When we consider the ways in which those alleles could have reached western Europe, where agriculture developed later, it's worth remembering that it only takes a few months to walk that entire distance. If that happened often, you'd expect genes to have spread far more rapidly than they actually did. The problem is that would-be long-distance travelers quickly encountered other groups that spoke a different language and had uncongenial customs. Some were enemies, and all were suspicious of strangers. Passing through other groups was very difficult, so long-distance land travel was almost impossible.

In *The Third Chimpanzee*, Jared Diamond describes this pattern in highland New Guinea, one of the last places in the world to make contact with outsiders. He wrote, "When I was living among Elopi tribespeople in west New Guinea and wanted to cross the territory of the neighboring Fayu tribe in order to reach a nearby mountain, the Elopis explained to me matter-of-factly that the Fayus would kill me if I tried. From a New Guinea perspective, it seemed so perfectly natural and self-explanatory. Of course the Fayus will kill any trespasser: you surely don't think they're so stupid that they'd admit strangers to their territory? Strangers would just hunt their game animals, molest their women, introduce diseases, and reconnoiter the terrain in order to stage a raid later."[9] Before outside contact, New Guinea highlanders spent their entire lives within a few miles of their villages, and as far as we know, none had ever seen the sea, which was just 100 miles away. It seems likely the whole world was like this in prehistory.

## HISTORICAL PATTERNS OF GENE FLOW

Trade gave people reasons to surmount or bypass those barriers to travel, and it has played an important role in facilitating human gene flow. Although its impact on gene flow was probably not as dramatic as that resulting from conquest and colonization, it was important, particularly after the development of sailing ships. Sailors and barmaids, like traveling salesmen and farmers' daughters, have played a crucial role in recent human evolution.

Trade—more exactly, traders—spread new alleles along every line of communication: along sea routes, up navigable rivers, between villages and market towns. In early times it often took the form of semi-military campaigns, as when Egypt sent trade expeditions to Punt in northern Ethiopia/Eritrea in order to obtain gold, slaves, ebony, and ivory, or when Mesopotamian kings like Sargon of Akkad sought cedarwood in Lebanon.

Trade connected widely separated civilizations—some of them, some of the time—nearly as far back as we have records. There is evidence of trade between the Indus civilization and Mesopotamia during the Akkadian Empire more than 4,000 years ago—Mesopotamian cereals and wool in exchange for wood and ivory from "Meluhha," thought to be the Indus civilization. We have found Indus cylinder seals in Ur and Babylon, and references to a village made up of traders from Meluhha, who eventually became one more component of the local ethnic mix. This is exactly the sort of contact that could have played a significant role in the transmission of new beneficial alleles: The peoples of the Indus civilization probably faced similar selective pressures to those seen in Mesopotamia, and they were far enough away that gene flow through local marriage would have taken a long time to occur. However, the contact happened early in history, so that any good alleles introduced to the Middle East in this way have had 160 generations in which to spread.

## COLONIZATION

One major pattern of migration that affected gene flow was the seeding of colonies along the coasts of the Mediterranean and Black seas by peoples from the eastern end of the Mediterranean. Even as far back as the late Bronze Age, it was easier and cheaper to travel long distances by sea, not least because you could avoid having to fight your way through already-established peoples. Long-distance trade helped pave the way for these colonies. Sometimes they began as trading posts, and trade must have been the basis for the seafaring techniques and geographical knowledge that made the founding expeditions possible.

The colonizers—Etruscans, Greeks, and Phoenicians—came from the Middle East or areas heavily settled by early Middle Eastern farmers. They must have carried many of the new alleles that were adaptive in the context of agriculture, and their voyages probably spread many copies of those alleles to the western Mediterranean.

The Phoenicians, a people living in cities along the coast of what is now Lebanon, traded over much of the Mediterranean. Their colonies started out as anchorages and trading posts along their trade routes, but eventually some of them grew into substantial towns. The largest and most important of these colonies was Carthage (Kart-Hadasht) in Tunisia, which eventually became Rome's great rival. Many of those towns still exist today: Palermo and Marsala in Sicily, Cagliari in Sardinia, Tangier in Morocco, and Cadiz and Cartagena in Spain. Phoenician colonies appear to have spread one of the more common versions of beta-thalassemia (a genetic defense against malaria) around the western Mediterranean. Most likely this allele originated in North Africa.[10]

The Greeks colonized on a larger scale: A single city, Miletus, founded ninety colonies. Some were founded for commercial advantage, some as a refuge for a losing side, and others as a means of getting rid of surplus population. The Greeks founded many colonies in Sicily and southern Italy, which became known as "Magna Graecia." They include such modern cities as Syracuse and Naples. There were also many Greek colonies around the Black Sea, some (like Marseille) in southern France and others as far away as Spain and Libya. It looks as if the Greeks spread at least two characteristic malaria defenses of their own, a different version of beta-thalassemia and a form of G6PD deficiency. (Malaria defenses are probably not the only adaptive alleles that Phoenicians or Greeks transmitted—they are just the ones that have been well studied so far.) Since the recent whole-genome surveys show that many genes have been under recent selection and risen to high frequencies, apparently mostly in response to the new conditions that accompanied agriculture, we expect that these colonizations transmitted a number of adaptive alleles.

Recent genetic studies have confirmed a third major colonization in which an eastern group, this time from Turkey, colonized northwestern Italy: the Etruscans. They were a somewhat mysterious people who spoke a non-Indo-European language that we have not yet deciphered. The Etruscans had tremendous influence on Roman art, architecture, and religion. The question of Etruscan origins had long been controversial, with most archaeologists arguing that the culture developed in Italy, although some ancient sources, such as Herodotus, said that it originated in Lydia, a region on the western coast of Turkey. Recent work has shown that some populations in Tuscany have Near-Eastern mtDNA[11] and that some distinctive local *cattle* also have mtDNA characteristic of Middle Eastern breeds,[12] confirming an Anatolian origin.

The Etruscans added a healthy dose of Middle Eastern, agriculture-adapted alleles into the Roman mix. We have reason to suspect that those alleles shaped attitudes as well as affecting metabolism and disease resistance. Did they influence Rome's rise to power? It's possible.

## YOU HAVE BEEN IN AFGHANISTAN, I PERCEIVE

Military movements also let favorable alleles vault over long distances and geographical barriers. Alexander the Great furnished one of the more dramatic examples. In the course of a remarkable career of conquest (dying undefeated), he marched as far east as Pakistan.

In addition to settling Greeks over much of the Middle East, his legacy included Greek kingdoms that survived for several centuries in Afghanistan and Pakistan. Those kingdoms didn't just influence the artistic development of Buddhism; they also transmitted alleles. Today we see a few Greek Y chromosomes among the Pathans, the dominant ethnic group in Afghanistan.[13] Of course, we also see some Y chromosomes that are directly descended from Genghis Khan himself in the Pathans' despised neighbors, the Hazara.[14] Local marriage could never have spread genes as rapidly as that—but Genghis and Alexander could.

If regional Y-chromosome variants (which as far as we know have no special inherent fitness advantage) could spread that far, you can be sure that any advantageous mutation that had become common in Greece in Alexander's time did as well. Every such allele has had a good chance of becoming common in Afghanistan by the present day. This isn't quite as true for those Mongol alleles, since they've only had a third as long to spread as those of Alexander and his merry men. In these long-distance transfers, the earlier the connection, the more important. And in the same way, large population transfers have a greater effect than small ones.

## THE LOST TRIBES

Imperial politics sometimes played an important role in dispersing genes, often in highly unpleasant ways. Forced relocation of peoples has been a standard tactic during times of conflict—whether in ex-Yugoslavia or Chechnya in our day or in Assyria during ancient times. Tiglath Pileser III moved some 30,000 people from what is now northern Syria to the Zagros

Mountains in western Iran in 742 BC, Sargon II displaced about 100,000 Babylonians in 707 BC, and Sennacherib deported another 208,000 in 703 BC. One of the most notorious forced relocations was in 722 BC, when the Assyrians conquered the Northern Kingdom of Israel, destroying its capital and sending its population into exile.

These population transfers were intended not only to punish, but also to break up local elites and traditions, open up strategic areas for occupation, and provide the Assyrian state with labor and soldiers. The number of people forcibly removed from home over three centuries of these policies has been estimated at more than 4 million. Considering the enmity these actions provoked, they may have hastened the fall of Assyria—but they surely spread alleles over most of the Fertile Crescent.

There are other famous examples of forced relocations. The Babylonian Empire defeated Judah, the Southern Kingdom of Israel, in 586 BC and relocated some of the population to Mesopotamia. After the Persian Empire succeeded the Babylonians in 539 BC, Cyrus the Great allowed them to return.

## AN ARTHURIAN ROMANCE

The Sarmatians were steppe nomads from the southern Ukraine who spoke an Iranian language. The classical historian Cassius Dio said, "The Sarmatians were a savage uncivilized nation, . . . naturally warlike, and famous for painting their bodies to appear more terrible in the field of battle. They were known for their lewdness. . . . They generally lived on the mountains without any habitation except their chariots. . . . They lived upon plunder, and fed upon milk mixed with the blood of horses."[15]

They were famous for their heavy cavalry, who fought with lances, longswords, and bows. The Romans had fought them in AD 92 and knew their quality. In AD 175, Marcus Aurelius hired 8,000 Sarmatians into Roman service and sent 5,500 of them to northern Britain. At first they were attached to one of the Roman legions there, Legio VI Victrix, but when their twenty years of service were up, they were settled in a permanent military colony in Lancashire. Apparently they never went home: The colony is still mentioned almost 250 years later.

Imagine that one of those selected alleles we see in the HapMap scans originated far to the east of Britain—perhaps as far east as Kazakhstan—some thousands of years ago, possibly in the Andronovo culture. Then, suppose the allele had a large selective advantage, and by the time the Sarmatians were fighting for the Romans, it had become common among the Iranian-speaking steppe peoples, among whom it had diffused easily because of their characteristic horse-nomad mobility—but had not yet spread as far as western Europe. If limited to girl-next-door diffusion, it would have required several millennia to reach Britain.

That Sarmatian military colony, however, could have introduced several thousand copies of that hypothetical allele into Lancashire. The Sarmatian cavalrymen were paid well and surely could have managed to raise at least as many children as the average Briton. Starting with an original gene frequency of 0.1 percent in England in the year 175, that hypothetical allele could have a high frequency in the English population by the present time. Trade and war would ensure that the new allele spread effectively over Great Britain, and there was undoubtedly plenty of both, especially war.

Those Sarmatians may have spread ideas as well. They had some very interesting religious beliefs and legends, some of which are preserved among the Ossetians, their descendants in the Caucasus. A number of those legends sound awfully familiar—in particular, the story of a dying warrior who demands that his best friend destroy his sword by tossing it into a lake rather than allowing it to be captured by enemies. The friend can't see throwing away such a beautiful weapon, and twice pretends to have done so—but the sword-bearer, hearing his account, somehow knows that he has not. On the third try, the sword is thrown into a lake and is caught by a woman's hand coming out of the water.

So it may be that the Sarmatians introduced key elements of the Arthur mythos, the Matter of Britain, the subject of books and poems and movies for hundreds of years. A good story can go a long way, as can a good allele. There is a real similarity. A slight contact can transmit an idea, if it falls on fertile ground—if people like the idea and repeat it. In the same way, a long-forgotten Roman transfer of troops may have played a key role in the genetic history of Britain. A few copies of a favorable allele can increase tremendously, given time: The average Englishman has only a dab of Sarmatian ancestry, but might be mostly Sarmatian in a key gene or two.

## BLUE EYES

A strong state fosters gene flow within itself through trade, free movement, and sometimes forced movement. At the same time, it tends to limit gene flow from outside, particularly when it upholds its borders with military force. When such a state disintegrates, there can be a massive movement of peoples—in part because the borders are no longer defended, but even more in those cases in which the inhabitants have lost their military habits over the course of a long imperial peace. In classical times, one obvious sign of that kind of relative domestic tranquility was unwalled cities.

The fall of the Roman Empire followed this pattern. As the state weakened, many ethnic groups entered. In the early days they were often mercenary soldiers, but eventually some came as plundering bands. Those invaders also carried new alleles. We will discuss the Vandals, one of the most spectacular of these groups. We think they may have played a part in spreading a particular far-flung new allele, the one responsible for blue and green eyes.

Blue eyes are common in Europeans and their descendants and are found to some extent in adjacent populations, but they are essentially nonexistent in most of the world. Some 10,000 years ago there seems to have been no such thing. There are other shades of eye color, and other genes have some influence, but most of the story boils down to a single new allele of the gene named OCA2 (for oculocutaneous albinism II, also discussed in Chapter 4). To be exact, blue eyes are caused by a change in a DNA sequence that regulates the expression of OCA2, a sequence that is embedded in HERC2, the gene next to OCA2.[16] That allele accounts for 75 percent of the variation in eye color in Europe. It's the third longest haplotype in Europeans and therefore can't be very old: Analysis of the unshuffled region associated with OCA2 suggests that it originated about 6,000 to 10,000 years ago. Blue eyes are most common in northern Europe, centered around the Baltic. The simplest assumption

A member of the Tuaregs, nomadic Berbers of the Sahara

is that the allele originated in the center of the region, where its frequency is very high today, so our best guess is that it first occurred in a Lithuanian village about 6,000 years ago.

Clearly the spread of this allele involved more than just men marrying the girl next door. You can see light eyes in Berbers from the Atlas Mountains of Morocco, and even in Tuaregs living in the Sahara. Light eyes are fairly common in Kurds living in the Zagros Mountains along the border between Iraq and Iran, and we can find people with light eyes as far away as Afghanistan, over 3,000 miles from Vilnius. In order to explain these patterns, we're going to have to make a little excursion through history. Again, a gene is the center of attention rather than battles and kings.

Light-eyed Afghan girl

Start with the Berbers. Clearly the OCA2 allele didn't get there solely by peasants marrying their neighbors, since the Mediterranean Sea blocks that kind of connection. It could be that blue eyes in Morocco are caused by some local mutation, not the same one that has spread over Europe. But if it is the same mutation, as seems likely, since we have found the same OCA2 haplotype around the Mediterranean, the prime candidates for bringing that gene to North Africa are barbarians and pirates.

Barbarians came first. During the twilight of the Roman Empire, whole tribes, mostly some flavor of German, began wandering across the borders. The Vandals were among the most troublesome. They are thought to have originated in southern Sweden and had moved to Silesia (now eastern Poland) by 120 BC. By the third century, they had moved on to western Romania and Hungary, then into Roman territory. Around AD 400 they began moving westward (along with their allies, the Alans and Suebians), crossing the frozen Rhine on the last day of 406. They went on to devastate France, plundering their way through Aquitaine and crossing the Pyrenees in late AD 409. The Alans (an Iranian-speaking people off the steppe) set up a kingdom in Portugal, while the Suebians settled down in Galicia.

Facing pressure from the Visigoths (yet another medieval gang, which had already crushed the Alan kingdom), the Vandals and the surviving Alans, some 80,000 souls altogether, crossed from Spain into Africa in 429. This Vandal kingdom dominated the western Mediterranean with its stolen fleet, extorting tribute from seaports and sending out raiding parties every year. The Vandals sacked Rome in 435, ensuring that their name would live in infamy. Finally, in 533, Justinian, the emperor of the remaining eastern half of the Roman Empire, sent a force to attack the Vandals. Led by Belisarius, the best general of the age and one of the best of all time, the Imperials landed in Carthage and made short work of the Vandals—some of whom, when finally defeated, may have blended into the countryside. As Edward Gibbon wrote, "When every resource, either of force or perfidy, was exhausted, Stoza, with some desperate Vandals, retired to the wilds of Mauritania, obtained the daughter of a Barbarian prince, and eluded the pursuit of his enemies, by the report of his death."[17]

Seems like a lot of trouble just to inject a few thousand copies of the new OCA2 allele into the Rif.

Later came pirates. From 1500 to 1800, Muslim corsairs captured and enslaved many Europeans: by one estimate, more than a million. Mainly they took captives from the Mediterranean coasts of Italy and Spain, but occasionally they ranged farther afield, raiding Cornwall, Ireland, and even Iceland. Most male slaves were worked to death and made little contribution to the gene pool, but women could and often did end up in harems. You see a similar pattern with the slaves the Arabs took from sub-Saharan Africa: About 5 percent of the maternal ancestry of Arabs in the Middle East is African, judging from mtDNA, but you see very few African Y chromosomes there.

We're not suggesting that Berbers have a lot of European ancestry: Judging from Y-chromosome and mtDNA data, that does not appear to be the case. The point is that even a moderate degree of admixture can introduce many copies of a beneficial allele, and over time that allele can become common. The earlier the introduction, and the more copies introduced, the more effective this process is. If we had to guess, we'd say that the blue-eyed variant of OCA2 found in Berbers was probably introduced by the Vandals, but it may have happened earlier and involved population movements that we are not aware of.

# 6
# EXPANSIONS

History is full of examples of human groups expanding at the expense of their neighbors. Anatomically modern humans expanded and replaced archaic humans, the Bantu expanded at the expense of the Bushmen and other peoples, and Turks and Mongols pushed aside the Iranian-speaking peoples who had previously occupied the steppes of central Asia. In many of those cases, there was some degree of admixture, but replacement dominated. We could cite dozens of other examples. Surely the most obvious question is *why* those groups expanded.

In some cases, sheer chance may have played an important role—perhaps some key battle was lost for want of a horseshoe nail. More often, the successful group had some kind of advantage that drove their expansion. Anatomically modern humans probably had new and improved technologies like projectile weapons as well as more sophisticated language. The Bantu had iron tools and a set of domesticated plants adapted to Africa, which together gave them a pretty powerful advantage over the hunter-gatherers they encountered. The Turco-Mongol advantage over the steppe Iranians is less obvious, but they may have had stronger political organizations.

The general assumption is that the winning advantage is cultural—that is to say, learned. Weapons, tactics, political organization, methods of agriculture: all learned. The expansion of modern humans is the exception to the rule—most observers suspect that biological differences were the root cause of their advantage. Biological advantages are particularly potent because they last: Archaic humans such as Neanderthals may have been able to copy some of the cultural attributes of modern humans (exemplified by the Châtelperronian toolkit), but they couldn't *become* modern humans, couldn't copy or acquire abilities that were consequences of modern human biology. So being an anatomically modern human was an enduring advantage, and thus genetics can explain a replacement process that seems to have taken about 20,000 years (from the original trek out of Africa to the last Neanderthals).

The assumption that more recent expansions are all driven by cultural factors is based on the notion that modern humans everywhere have essentially the same abilities. That's a logical consequence of human evolutionary stasis: If humans have not undergone a significant amount of biological change since the expansion out of Africa, then people everywhere would have essentially the same potentials, and no group would have a biological advantage over its neighbors. But as we never tire of pointing out, there *has* been significant biological change during that period—tremendous amounts of change, particularly in those populations that have practiced agriculture for a long time. Therefore, the biological equality of human races and ethnic groups is not inevitable: In fact it's about as likely as a fistful of silver dollars all landing on edge when dropped. There are important, well-understood examples of human biological inequality: Some populations can (on average) deal far more effectively with certain situations than others.

New alleles that have undergone selective sweeps under agriculture show up randomly—so they might occur more in some groups than in others by sheer chance. We know that they become common first where they're favored—in the new agricultural ecology, which some peoples experienced before others. Thus, the early adopters got a head start on new adaptive genes. And the way these advantageous alleles propagate, people along lines of communication are likely to pick up more of them than people who live off the beaten path.

Early adopters ought to be better at agriculture than latecomers: They should be better adjusted to the new diet, tougher against the new diseases, and better at tolerating crowding and hierarchy.

Those new advantages all had to increase individual fitness, but their effects at the level of the tribe or ethnic group varied. Some aided individual survival but didn't have much effect at the group level. For example, a mutation that protected against an infectious disease wouldn't have had much effect on overall population size if food shortages were the major limiting factor. However, a new allele that allowed its bearers to digest a new food more effectively might well have increased the size of that same group. A mutation that helped its bearers compete with other humans without conferring any other advantage, one that changed the winners without bringing any more money into the game, probably wouldn't have helped the group; it might even have weakened it.

If a group happened to acquire one (or a few) of those mutations that increase *group* fitness as well as individual fitness, it would have had a real advantage over its neighbors. Its population would have expanded. Tribes and bands fight, of course—they always have. War goes back well before the birth of civilization.[1] But populations with a biological advantage, more often than not, should have won the wars. They would have been able to generate more young warriors than their neighbors. They would have been able to afford to fight more often and recover faster from defeat. If the expanding group's success depended upon some improved tactic or weapon, the defenders could have copied it. But they couldn't copy a gene. It's hard to fight biological superiority, and expansions based on such superiority could have gone on far longer than ones based upon cultural advantages, which are ephemeral.

## THE COLUMBIAN EXPLOSION

Other writers have discussed the crucial role of epidemic Eurasian and African diseases in the European expansion into the Americas, but most shy away from clearly stating that this was driven by underlying biological differences—differences that conferred a practical advantage, a kind of superiority, in this particular situation. But there is plenty of evidence that these biological differences existed. When Europeans launched their first ships to the "New World," their diseases came along for the ride, and the Amerindians simply lacked the biological defenses they would have needed to withstand that onslaught.

The Amerindians migrated from Northeast Asia some 15,000 years ago. They did not carry with them crowd diseases that arose after the birth of agriculture, nor did they carry the genetic defenses that later developed against those diseases. Since their path to the New World went through frigid landscapes like Siberia and Alaska, they left behind some of the ancient infectious diseases that were vectorborne or had complex life cycles—malaria and Guinea worm, for example. The world they entered had never before been settled by hominids or great apes, so there were few local pathogens preadapted to humans. Many of the infectious diseases found in the Old World are thought to have originated in domesticated animals, but this does not seem to have been an important factor in the Americas.

Although Amerindians did develop agriculture independently—a very effective agriculture that included some of the world's most important crops, such as maize and potatoes—they domesticated few animals, mostly because they had already wiped out most of the species suited to domestication. That happened whenever modern humans, who were competent hunters, entered a land that had never known any kind of humans before, one in which none of the large animals had had a chance to adapt to humans. It happened in Australia, New Zealand, and Madagascar as well as the New World.

So the selective pressures favoring disease resistance were weaker among the Amerindians than among the inhabitants of the Old World—possibly weaker than that experienced by any of our ancestors for millions of years.

One sign of this reduced disease pressure is the unusual distribution of HLA alleles among Amerindians. The HLA system (for human leukocyte antigen) is a group of genes that encode proteins expressed on the outer surfaces of cells. The immune system uses them to distinguish self from nonself, so they play an important part in rejection of transplanted organs. But their most important role is in infectious disease. There they present protein fragments from pathogenic organisms such as bacteria to immune system cells that then attack the pathogen. In addition, when a virus infects a cell, HLA molecules display viral proteins on the outside of the cell, so that those infected cells can be destroyed by the immune system.

HLA genes are among the most variable of all genes. There are ten or more major variants of each HLA gene, and most have more than 100 variants. Because these genes are so variable, any two humans (other than identical twins) are almost certain to have a different set of them. Because the alleles are codominant, having different HLA alleles expands the range of pathogens that our immune systems can deal with. Natural selection therefore favors diversification of the HLA genes, and some alleles, though rare, have been preserved for a long time. In fact, some are 30 million years old, considerably older than *Homo sapiens*. That is to say, there are HLA alleles in humans that are more similar to an allele in an orangutan than to other human alleles at that locus. Selection favoring HLA diversity—a selective pressure stemming from infectious disease—has existed more or less continuously for tens of millions of years. This is why even small populations in the Old World retain high HLA diversity.

But Amerindians didn't have that diversity. Many tribes have a single HLA allele with a frequency of over 50 percent.[2]

Different tribes have different predominant alleles: It seems as if the frequencies of HLA alleles have drifted randomly in the New World, which hasn't happened since the Miocene in the Old World. A careful analysis of global HLA diversity confirms continuing diversifying selection on HLA in most human populations but finds no evidence of any selection at all favoring diversity in HLA among Amerindians.[3]

And if infectious disease was so unimportant among Amerindians, selection most likely favored *weaker* immune systems, because people with weaker immune systems would be better able to avoid autoimmune disorders, in which the immune system misfires and attacks some organ or tissue. Type 1 diabetes, in which the immune system attacks the pancreatic cells that make insulin, and multiple sclerosis, where it attacks the myelin sheaths of the central nervous system, are well-known examples—both are rare among Amerindians. A less vigorous immune system would have been an advantage under those conditions.

So, there is every reason to think that the inhabitants of the Americas were not just behind the immunological times: While the Old Worlders were experiencing intense selection for increased resistance to infectious disease, the Amerindians were actually becoming more vulnerable. They were adapted to the existing circumstances, but not to the coming collision with the Old World.

These long-term differences in selection pressures had dramatic consequences when Columbus brought the Old and New Worlds into regular contact. Eurasian infectious diseases such as smallpox, measles, diphtheria, whooping cough, leprosy, and bubonic plague were introduced to the Americas in short order. In tropical and subtropical areas, they were eventually joined by yellow fever, dengue fever, falciparum malaria, lymphatic filariasis, schistosomiasis, and onchocerciasis (river blindness), most of which came from Africa. Relatively few pathogens, however, traveled in the other direction, from the Americas to the Old World. Syphilis[4] and tungiasis (a flea that burrows into the skin) are the only human pathogens we are aware of that originated in the New World and have spread to the Old, although there may be others we haven't yet recognized—for example, some epidemiologists suspect that rheumatoid arthritis is caused by a cryptic New World pathogen.

Suddenly exposed to this avalanche of unfamiliar infectious diseases, Amerindians were devastated. Some estimates suggest that the indigenous population of the Americas dropped by more than 90 percent over a few centuries, with almost all of the loss due to infectious diseases.[5]

This Amerindian vulnerability was a primary reason for European success in the Americas. Epidemic disease, particularly smallpox, interfered with armed resistance by Amerindians and thus played an important part in the early Spanish conquests. In Mexico, where Hernán Cortés and his troops had made the Aztec emperor their puppet, the Aztecs rose against them, killing Moctezuma II and two-thirds of the Spanish force in the famous "Noche Triste." The Aztecs probably would have utterly destroyed the invaders, were it not for the smallpox epidemic under way at the same time. The leader of the Aztec defense died in the epidemic, and Cortés and his men conquered the Aztec Empire.

It is hard to see how Cortés could have won without those microscopic allies, since he was trying to conquer an empire of millions with a few hundred men. Moreover, major Indian polities such as the Mayan city-states were still intact after the defeat of the Aztecs, and the Spaniards might well have lost control except for the series of epidemics that followed. Francisco Pizarro's conquest of the Incan Empire was also aided by a smallpox epidemic. It killed the emperor and his heir, causing a convenient succession struggle. Considering that Pizarro was invading another empire of millions with only 168 soldiers, it's obvious that he needed all the help he could get.

Amerindian vulnerability to infectious disease shaped history again and again. The first Spanish attempts at colonization in the West Indies were actually jeopardized by it, since the Taino and Arawak peoples diminished so rapidly (they were almost gone by 1530) that the Spanish were left without a labor force. Inhabitants of those Caribbean islands had been even more isolated and shielded from disease than the Amerindians of the mainland, and so were even more vulnerable.

The Pilgrims' first settlement was on land that already had been cleared by an Indian tribe that had been ravaged by some plague (possibly smallpox) just three years earlier. Squanto, the Indian who taught the Pilgrims survival skills, seems to have been one of the few survivors of that tribe. The later Puritan settlement of New England was also furthered by devastating epidemics among the Amerindians, while Jamestown's safety was only secured when epidemic disease had weakened the local tribes.

The Amerindians survived best in the highlands, where they could avoid most of the new African diseases.[6] In fact, in the Altiplano of South America, which has an altitude of over 11,000 feet, local Amerindians had a compensating advantage over the Spaniards because they were better adapted to the thin air.

Europeans had multiple advantages, of course. They had superior weapons and tactics, honed over thousands of years of organized conflict. Their form of warfare was more realistic and less ritualistic, at least in comparison to that of the Aztecs, who had "flower wars"—wars aimed at acquiring captives for sacrifice rather than achieving decisive results—with neighboring city-states such as Tlaxcala. Europeans had a varied and useful set of domesticated animals to use for food, raw materials such as wool and leather, and transportation. They had advanced metallurgy (iron and steel) and large sailing ships. They were the heirs of literate cultures going back thousands of years, and although there were exceptions, such as Pizarro, who never learned to read or write, many of the early explorers and settlers—the Puritans, for example—were highly literate and well educated.

The most complex Amerindian civilizations in 1492 were similar to civilizations found in the Middle East 3,000 to 4,000 years earlier—so the Europeans were, in a sense, invaders from the future.

The European advantage in disease resistance was particularly important because those early attempts at conquest and colonization were marginal. Shipping men and equipment across the Atlantic Ocean presented huge logistical difficulties. European military expeditions to the New World were tiny and poorly supplied. The successes of the conquistadors are reminiscent of ridiculous action movies in which one man defeats a small army—and that's a lot harder to do with an arquebus than an Uzi. Early colonization efforts often teetered on the edge of disaster, as when half the Pilgrims died in their first winter, or when most of the settlers in Jamestown starved to death in the winter of 1609.

Epidemic disease didn't just grease the skids for the initial conquests: It reduced Amerindian populations and made later revolts far weaker than they would have been otherwise. If they had not died of disease, the Amerindians would have had time to copy and use many European military innovations in the second or third round of fighting.

We know a lot about the genetic basis of resistance to malaria, but relatively little about the genetic basis of European resistance to diseases like smallpox, although there are some hints. As we have said before, there is plenty of evidence for selection acting recently on many genes involved with disease defense, but in most cases we don't know the biochemical details—for example, which particular infectious organism a particular selected allele defended against. We suspect that delta CCR5 (for chemokine receptor 5), a common mutation among northern Europeans, protects against smallpox, but since smallpox is dangerous to work with and now exists only in a couple of genetic repositories, it's hard to be sure.[7] Some recessive genetic diseases that are common in Europe and the Middle East also probably have conferred resistance to some infectious diseases: That list would include cystic fibrosis, alpha-1-antitrypsin deficiency, familial Mediterranean fever, connexin-26 deafness, and hemochromatosis. All are nonexistent in Amerindians, discounting recent admixture.

There is another way we may be able to detect some of the alleles that helped protect Europeans from infectious diseases that devastated the Amerindians: admixture studies. Many of the present inhabitants of Latin America are descended from both Europeans and Amerindians, along with a smaller amount of African ancestry. In the absence of natural selection, you'd expect that admixture would be the same at each gene, allowing for chance—if 40 percent of overall ancestry was European, then 40 percent of the copies of each gene would also be European. But as we pointed out in Chapter 2, this no longer remains the case if a particular allele of some gene causes increased reproductive fitness. For example, if the European allele of some gene conferred substantial protection against smallpox, the average inhabitant of Mexico might be considerably more European at that locus than average—even though the admixed population is no more than 500 years old. In other words, there could be adaptive introgression of the European version of that gene. Some Amerindian or African alleles might also have advantages: But considering the relatively short time available for introgression (at most twenty generations) and the sheer destructive power of smallpox, a European smallpox defense seems one of the more likely candidates for detectable introgression. There is evidence of such an unusually European chunk of the genome in at least one Mexican American population.[8] It is remarkable that the same principles apply to Neanderthals and conquistadors.

We do know a bit about immunological differences between Amerindians and other peoples. We know that the Yanomamo (a much-studied group of Amerindians in the backwoods of Venezuela) tend to produce high levels of antibodies against tuberculosis antigens rather than the more effective cell-mediated responses seen in Europeans. Even though tuberculosis is common among the Yanomamo, few individuals have a positive response to a tuberculin test. This is important, because most Old Worlders exposed to tuberculosis mount an effective immune response (which causes a positive response on a tuberculin test) and avoid symptoms. Only a minority develop active disease.

The Yanomamo also have extremely high levels of Immunoglobulin E (a molecule involved in defense against parasitic worms)—much higher than levels seen in Europeans with the same level of infestation. Pre-Columbian Amerindians were subject to some parasitic worm infections but relatively few bacterial or viral diseases, and it may be that natural selection adjusted their immune system to face those threats.

We know more about the practical consequences of these genetic differences than we do about their biochemical details, in part because of historical accounts of the relative impact of infectious diseases in European and Amerindian populations, but also because of well-documented epidemics during the era of scientific medicine—the past 100 years or so.

Even during the twentieth century, first contacts between Amerindians and people of European descent killed one-third to one-half of the natives in the first five years unless there was high-quality medical care available.[9] This was the case during a period in which some of the worst Eurasian diseases (smallpox, bubonic plague, and typhus) were no longer major threats. For example, of the 800 Surui contacted in 1980 in Brazil, 600 had died by 1986, most of tuberculosis.

Judging from historical accounts, the fatality rate of smallpox was much higher among Amerindians than among Europeans. Roughly 30 percent of the Europeans who were infected died, whereas for the Amerindians, the fatality rate sometimes reached 90 percent. For example, in an epidemic in 1827, smallpox spared only 125 out of 1,600 Mandan Indians in what later became North Dakota.

Some historians have argued that a virulent epidemic hitting an epidemiologically inexperienced population would be especially damaging because it kills adults rather than children. It takes a long time and a lot of investment to produce an adult, so they are hard to replace. Since adults do most of the productive labor and produce most of the food, this matters. A population can survive a disease that kills 20 percent of the population in childhood more easily than it can survive one that kills 20 percent of the population in early adulthood. This effect may have increased the impact of the first wave of epidemics in the New World. Along the same lines, an epidemic that sickens nearly everybody may leave too few caretakers to nurse those who could survive if they were fed and kept warm. However, this effect can't explain why infectious diseases kept hitting Amerindians harder than Europeans in epidemic after epidemic over hundreds of years.

Although factors such as a paucity of domesticated animals decreased the probability that the Amerindians would develop really potent infectious diseases of their own, it must have been possible. In principle, they might have had their own equivalents of smallpox and malaria. But if such potent diseases had been brought back to the Old World by European explorers, civilization would have fallen and you wouldn't be reading this right now. Some have also said that cultural inexperience might have worsened these epidemics, as when stricken Indians ran from epidemics (thus spreading the disease further) or tried various ineffective therapies. But Europeans ran from epidemics, too (as in Boccaccio's *Decameron*), and European medicine in those days was generally useless. (Charles II's doctors in the 1600s treated his fits with bleeding, cupping, emetics, laxatives, enemas, blistering plasters, Spanish Fly, and more bleeding, then plastered the soles of his feet with tar and pigeon dung, gave him a bezoar [a concretion found in the stomach of a goat, thought to neutralize any poison], and followed this up with more bleeding. None of it worked.)[10]

But wasn't Spanish oppression of the Amerindians the main factor reducing their numbers? We think not. Of course, the Spanish did oppress the Amerindians, but they were hoping to become the new lords of those lands, rather than farming the land themselves. Lords need serfs—live serfs. Spanish demands for labor and food must have made the situation worse, but depopulation raced far ahead of Spanish administrative control. For example, when Hernando de Soto explored the American South in 1539, he found many fair-sized towns, but also ghost towns that had been recently abandoned. Old World diseases (likely smallpox) had gotten there first, just as smallpox had reached Peru before Pizarro.

Furthermore, the Spanish began to conquer the Philippines in the sixteenth century, and there's no sign that they caused a population collapse there.[11] Whenever Europeans made contact with a long-isolated people, whether Amerindians or Australian Aborigines or Polynesians, there was a population crash. When Europeans conquered peoples who had already had extensive contact with other Old World populations, as the British did in India or the Dutch in Indonesia, there was no crash. As Charles Darwin said, "Wherever the European has trod, death seems to pursue the aboriginal. We may look to the wide extent of the Americas, the Cape of Good Hope, and Australia, and we find the same result."[12]

Those who refuse to acknowledge the crucial role of biological differences in the European conquest and settlement of the Americas are rejecting Darwinian evolution. Thousands of years of high disease load among Eurasian agriculturalists *had* to select for increased disease resistance. This isn't particularly unusual or unorthodox, yet many who claim to accept the idea of natural selection reject most of the obvious implications of the theory when it is applied to humans.

Of course, as the Duke of Wellington said, two can play at this game. When the Europeans tried to conquer and settle sub-Saharan Africa, the shoe was on the other foot.

## HEART OF DARKNESS

Europeans had long been aware of the lands south of the Sahara, but in the fifteenth century they knew very little about them. The information they did have was largely an inheritance from classical civilization. Hanno the Navigator (of Carthage) had explored the west coast of Africa, and Herodotus tells us of an earlier Phoenician expedition, sent out by the Pharaoh Necho around 600 BC, that seems to have circumnavigated the continent. Somehow, the Greeks acquired some information about central Africa, including facts about the Pygmies: Aristotle said, "These birds migrate from the steppes of Scythia to the marshlands south of Egypt where the Nile has its source. And it is here, by the way, that they are said to fight with the pygmies; and the story is not fabulous, but there is in reality a race of dwarfish men."[13] Another interesting hint is a 2,500-year-old frieze in Persepolis, the capital of the Persian Empire, which shows peoples from many lands bringing tribute. One panel shows a Pygmy with an okapi, a deep-forest relative of the giraffe that was only rediscovered by Europeans in 1901.

Sub-Saharan Africa may actually have been easier to reach and explore in classical times than it is today. The Sahara had not yet become as bone-dry as it is now: Horses could cross the desert until Roman times, and there are old, shallow wells in places where the water table is now thousands of feet below the surface. It's also possible—although this is speculation—that falciparum malaria wasn't as widespread in Africa in classical times as it is today. If true, African exploration might have been safer in those days.

When modern Europeans (Portuguese in the beginning) did start showing up along the coast of West Africa around 1500, they found wealth (mainly in gold and slaves), but they also faced incredible disease risks. The king of Portugal sent an expedition of eight up the Gambia River in about 1500: One came back alive. João de Barros, a Portuguese historian of the sixteenth century, said, "But it seems that for our sins, or for some inscrutable judgment of God, in all the entrances of this great Ethiopia we navigate along, He has placed a striking angel with a flaming sword of deadly fevers, who prevents us from penetrating into the interior to the springs of this garden, whence proceed these rivers of gold that flow to the sea in so many parts of our conquest."[14] Europeans typically bought slaves in coastal outposts or islands: Going inland to seize them was just too dangerous to their health. Arab slavers ranged further, but many Arabs had genetic malaria defenses such as alpha-thalassemia, and many were part African.

These difficulties persisted for centuries. British soldiers stationed on the Gold Coast would lose half their numbers in a year. Early explorers did no better. Mungo Park began his second attempt at African exploration (in 1805) with a party of forty-five Europeans; only eleven were alive by the time they reached the Niger. He was eventually killed at the Bussa rapids—by Africans, not parasites—but when his son Thomas went in search of him, *he* died of fever before getting far. We presume you've heard of Dr. Livingston—Dr. David Livingston, that is, the nineteenth-century British medical missionary to central Africa. His wife died of malaria during their travels, and the doctor himself later died of malaria and dysentery. John Speke and Sir Richard Francis Burton, nineteenth-century British explorers, sought and eventually found the sources of the Nile—but both men fell ill of tropical diseases. Speke suffered greatly when a beetle crawled into his ear. He removed it with a knife, but he became temporarily deaf and later temporarily blind. Consider that these are the famous explorers, the ones who enjoyed some degree of success. What happened to the unlucky ones?

Europeans had a vast technological edge over most of the inhabitants of sub-Saharan Africa. In most ways (except for their use of iron tools), African technology and social organization were simpler than that of the Amerindians—at any rate simpler than the Andean and Mesoamerican civilizations. (Here we're speaking of the inhabitants of what has been called the "isolated zone," areas that had not been much influenced by Islamic civilization—especially west, central, and southern Africa.) Literacy, the wheel, sailing ships, and guns gave the Europeans a huge military advantage, but nothing came of it for hundreds of years, except in the far south, where a temperate climate allowed Dutch colonization.

In the 1800s, quinine became widely available, and that allowed Europeans to venture into interior Africa with moderate success, since falciparum malaria had been the deadliest of many African diseases. Later scientific advances controlled or eliminated a number of other local diseases, including yellow fever and sleeping sickness. This made possible the "scramble for Africa," in which European countries ranging from Great Britain to Italy conquered almost the entire continent. In these efforts, European military technology was a trump card. As Hilaire Belloc wrote in a poem, "Whatever happens / We have got / The Maxim gun /And they have not."

But Africa did not become another America: Africans were not displaced by Europeans. In order for limited numbers of colonists to become the predominant population, the locals must die off, and Africans didn't. Powerful tropical diseases, combined with the local biological defenses (evolved at vast cost), kept Africa African. As in the case of the Columbian expansion, recent human evolution played a key role in determining the victors.

## THE COWBOYS

The genetic advantages in the two examples we have just discussed—for the Europeans in America and for the Africans in Africa—were huge: Those lacking the required resistance to the infectious diseases in play were almost wiped out. There is no reason to think that differences in disease resistance were the only biological differences between Europeans and Amerindians, or the only advantage, but they must have had the largest impact. European colonization could not have prevailed without a huge edge. Africans, too, may have needed a large biological advantage to resist Europeans, considering their technological and social disadvantages. Although we can't be sure, it looks as if anatomically modern humans also needed a fairly large advantage during the Upper Paleolithic as they displaced the Neanderthals, since they had to outcompete populations of archaic humans that must have been better adapted to Eurasian climates than they were.

And yet, there must have been occasions in which smaller biological advantages were enough to drive a population expansion, particularly when those expanding didn't have to cross oceans. Again, we're not saying that *all* expansions had such causes, but some could have, and the enduring nature of biological advantage makes it a good candidate for the cause of particularly widespread and long-winded expansions.

One of the largest of all known expansions—the spread of the Indo-Europeans—was likely driven by the mutation that conferred lactose tolerance, one of the most strongly selected alleles that Europeans possess.

"Indo-European" refers to a family of related languages that have spread over western Eurasia, the Americas, and Australasia. In terms of numbers, it is the largest of all language families, with about 3 billion native speakers, half of the human race. The largest Indo-European languages are Spanish, English, Hindi, Portuguese, Bengali, Russian, German, Marathi, and French.

In these languages, basic words in a number of categories are recognizably similar. In each of them, many of the words for numerals from one to ten, body parts (head, heart, and foot), plants and animals (oak, wolf, bear), natural phenomena (air, snow, moon), and close relations (father, mother, daughter) ultimately derive from a common ancestral language. For example, the word for "three" is *treis* in Greek, *tres* in Latin, *drei* in German, *tri* in Russian, *tri* in Bengali, and *tre* in Tocharian A, an extinct language of central Asia.

These languages were first acknowledged as a family when various Europeans in India noted similarities between Indian languages and European languages, particularly in regard to their connections with Latin and Greek. It was then suggested that a wide swath of languages in Europe and India had a common origin, just as it had long been recognized that the Romance languages (Spanish, Portuguese, French, Italian, and Romanian) derived from Latin. Most of those early observations were not followed up, but after Sir William Jones, an eminent scholar and chief justice of India, mentioned the pattern in a lecture on Indian culture in 1786, people began to take the idea seriously. Many people studied Indo-European language over the next two centuries, and today it is the most successful theory in historical linguistics.

People of many races and ethnic groups speak an Indo-European tongue: There is nothing genetic about that. Chinese pilots talk to Japanese air-traffic controllers in English, for that matter. But there is every reason to believe that the ancestor of all of these languages was once spoken by a particular people, living in some particular region. They were relatively few in number, and the region they occupied was small compared to the lands inhabited by Indo-European speakers today. There were many other small ethnolinguistic groups in Eurasia in those days, but this group spread, while others did not. Perhaps there was something unusual about them.

## THE PROTO-INDO-EUROPEANS

What we know about the Proto-Indo-Europeans, as we call this group, is mostly derived from comparative linguistics, supplemented by archaeology.[15] We know that they were stock-raisers and grain farmers, probably depending more on their animals than on grain. They raised cattle and sheep, along with goats and pigs. The cow played a paramount role, both in daily life and in religion. They had domesticated the horse, and in fact may have been the first to do so.

The Proto-Indo-Europeans knew copper, and probably bronze, but not iron. They used silver and possibly gold. They had wheeled vehicles, probably carts pulled by oxen. Woolen textiles were produced by weaving. They made and drank mead.

Their system was patriarchal, with clans tracing descent through the male line. They were warlike, constantly raiding for cattle and revenge. They probably had egalitarian warrior brotherhoods made up of single young men, with difficult initiation rites. Those warriors sometimes acted as berserkers in battle, probably had a wolf as a totem, and often were not quite kept under control by older and wiser heads.

Proto-Indo-European society as a whole was divided into three orders: a clerical class that administered the sacrificial rites of a polytheistic religion, a warrior class, and herder-cultivators. This division of society shows up in far-flung parts of the Indo-European dispersal: Ancient India has *brahmanas*, *ksatriyas*, and *vaisyas*, while Rome had *flamines*, *milites*, and *quirites*. French linguist George Dumézil and others have argued that this "tripartition" plays a key role in the religion and mythology of the Indo-European peoples, as when Herodotus tells how the kingship of the Scythians was awarded to one of three brothers who could pick up a burning cup, an axe, and a plow with a yoke. The three orders were color-coded. Priests wore white, warriors wore red, and the common people were symbolized by blue or black.

The Indo-Europeans practiced epic poetry, a form that used stock phrases, some of which show up in poetry that has been preserved to the present day, such as the *Iliad* or the *Rig Veda*. When someone refers to "driving cattle," or "undying fame," or "immortal gods," they are not being very original. Some of the Proto-Indo-European myths seem to have involved a world-tree or a hero slaying a dragon.

What we *don't* know, at least with any precision, is when and where the Proto-Indo-Europeans lived. Comparative linguistics offers a few hints about the time in which they still lived as one people (and had not yet begun to spread out) through the identification of technologies they had or didn't have. The overall level of technology (for example, bronze but no iron) suggests that dispersal began in the early Bronze Age, perhaps around 3000 BC. It had definitely begun by 2500 BC, since a settled Indo-European state (the Hittite Empire) shows up in the historical record a few hundred years later. We also see other Indo-European languages, such as Luwian and Palaic, in areas adjacent to the Hittite homeland in central Turkey: They are clearly related to Hittite, but must have been differentiating for some time (several centuries, at least) before they appeared in the historical record.

It's fair to say that the problem of the location of the Indo-European homeland, called "the Urheimat" (German for "original homeland"), has been a subject of controversy—indeed, the question has had a tendency to drive men mad. Various fruitcakes have suggested Tibet, North Africa, the shores of the Pacific, and the North Pole. There's a distinct tendency for scholars to place the wellspring of the European peoples somewhere in their own backyard. So far, thank God, we haven't seen any American linguists try that.

The two most popular theories regarding Urheimat locations place it either in Anatolia (modern Turkey) or the grasslands of southern Russia. Anatolia is the origin in the British archaeologist and linguist Colin Renfrew's model: His idea is that the Indo-European languages were carried along by an expansion of early farmers out of the Middle East around 7000 BC. There certainly *was* such an expansion: There is plenty of archeological and genetic evidence for it. The question is whether that expansion spread Indo-European languages.[16]

That idea is powerful because of the great population expansion associated with farming; numbers usually bring the victory. The idea is even more powerful than Renfrew suggested, in fact, because those Anatolian farmers had already been farming for millennia when they began to expand into the Balkans. As early adopters, they must have already been somewhat better adapted to the agricultural way of life than the native Europeans, and so they almost certainly had biological strengths that Europeans could not duplicate through observation and learned behavior.

Unfortunately, Renfrew's theory also has many fatal weaknesses. Linguistic paleontology supports a far later common origin than would be possible if the Proto-Indo Europeans were part of that Middle East expansion to the northwest into Europe. For example, there are several words referring to wheeled vehicles that are shared among Indo-European languages, but wheeled vehicles simply don't go back as far as 7000 BC. Hittite shows clear signs of a strong non-Indo-European substratum, as if Hittite invaders imposed their language on some other group that was already present in Anatolia. This can't make sense for the zone of origin. Uralic languages (the language family containing Finnish and Hungarian) appear to have had extensive contact with early Indo-European, and they may share a common ancestry. Since the Finnish peoples lived in the forest zone of what is now Russia, this suggests that the Indo-Europeans did not originate in the Middle East.

The second, more popular explanation is the Kurgan hypothesis, originated by Marija Gimbutas. In the 1950s she identified the Kurgan people of the Pontic-Caspian steppe (the grasslands between the Black Sea and the Caspian Sea) as the Proto-Indo-Europeans. If she is correct, they were a pastoral people who went through a series of expansions, which probably took the form of military conquests. Gimbutas thought they were mounted warriors and that their advantage stemmed from their early domestication of the horse. The problem is that there is no evidence at all of mounted warriors in this time period: indeed, not for at least another 2,000 years. The earliest horse-drawn chariots also appear far too late to explain Indo-European military expansion. Moreover, there is reason to believe that the winners in military conquests usually set themselves up as a dominant elite rather than wiping out those they conquered.

There is also a strain of thought that argues that Indo-European expansion was gradual and peaceful, in definite contrast to the way in which humans act today and have acted over the course of recorded history. Perhaps Gimbutas was correct in identifying the Kurgans as the Proto-Indo-Europeans, but had their modus operandi all wrong.

## MILK AND THE KURGANS

Improved variants of the Kurgan hypothesis fit many facts, but what they don't do is explain *why* the Proto-Indo-Europeans expanded at the expense of neighboring peoples with similar technology. Effective use of the horse in warfare doesn't seem to have occurred early enough to explain Proto-Indo-European expansion—but even if it had, what would have stopped other peoples from rapidly acquiring horses and using them in the same way? The Plains Indians certainly managed to master light cavalry warfare in short order: Why couldn't non-Indo-European peoples have done the same?

Later empires succeeded in part thanks to a snowball effect: The larger they grew, the stronger they were, until they were stopped by geographic barriers or long lines of communication. Once the Romans unified Italy, they were hard to stop. But as far as we can tell, nothing like this happened in the Indo-European expansion. It was too early for that kind of imperial organization. There was no central command, no capital, no state. If a peripheral Indo-European tribe had a dustup with neighboring non-Indo-Europeans, it had to win on its own, more or less. At most they had local allies. In order to expand as much as they did, early Indo-Europeans must have had some kind of edge, and in order to expand again and again over millennia, they had to have an edge that was hard to copy.

To solve the mystery, let's start with what we know about the Proto-Indo-Europeans from the linguistic evidence. We know that the Indo-Europeans weren't especially skilled at grain agriculture or adapted to it, since they were primarily pastoralists. They were removed from the first centers of farming in the Middle East. We also know that that the Proto-Indo-Europeans were rather backward in the realms of technology and social complexity. Sumerians invented the wheel, writing, and arithmetic and had cities and extensive irrigation systems at a time when the Proto-Indo-Europeans had, at most, domesticated the horse.

We suggest that the advantage driving those Indo-European expansions was biological—a high frequency of the European lactose-tolerance mutation (the 13910-T allele). The usual story about lactose tolerance is that it's the result of a cultural innovation, the domestication of cattle. That innovation led to selection for a new mutation that extended lactase production into adulthood. But there's more to the story.

Initially, selection favored individual carriers of the lactose-tolerance mutation, but the mutation was rare and had little social effect. Cattle were used for plowing and pulling wagons, for their beef, and as a source of secondary products like leather and tallow. But when the lactase-persistence allele became common, so that a majority of the adult population could drink milk, a new kind of pastoralism became possible, one in which people kept cattle primarily for their milk rather than for their flesh. This change is very significant, because dairying is much more efficient than raising cattle for slaughter: It produces about five times as many calories per acre.[17] Dairying pastoralists produce more high-quality food on the same amount of land than nondairy pastoralists, so higher frequencies of lactose tolerance among Indo-Europeans would have caused the carrying capacity of the land to increase—for them.

Standard ecological theory indicates that when two similar populations use the same resources, the one with the greater carrying capacity always wins. In more familiar terms, the Proto-Indo-Europeans in our scenario could raise and feed more warriors on the same amount of land—and that is a recipe for expansion. The same basic idea is behind theories of the expansion of farming through local population growth (called *demic expansion*): Farming produces more food per acre, therefore farmers will outnumber foragers, and so farmers will expand at the expense of foragers.

Proto-Indo-Europeans probably were most competitive in areas where grain agriculture was marginal. In the steppe, the problem was limited rainfall. Since raising cattle there had been competitive with grain farming even before dairying arose, milk-drinking Indo-Europeans would have had an absolute advantage and should have spread rapidly over the steppe. In much of northern Europe, shorter growing seasons must have interfered with production of cereal crops such as wheat, particularly when agriculture was new there, as those crops had had little time to adapt to the local climate. Eventually, other cereal crops, such as oats and rye that could do well in those climates, were developed—probably by accident, starting as weeds in wheat or barley fields. But that happened in the Bronze Age, long after the introduction of farming. Dairying may have been more productive than grain farming in northern Europe during the late Neolithic. Even if it was not, it may have been close enough to let other advantages of that pastoral way of life tip the scales. It seems clear that the Proto-Indo-European form of pastoralism did have other advantages in intergroup competition.

As the Proto-Indo-Europeans became dairymen, they should have come to rely more and more on their cattle and less on grain farming. As that happened, they would have become mobile, which is a military advantage, especially against farmers. Farmers have homes and villages that they must defend, whereas pastoralists can fight at a time and place of their choosing. Herodotus tells us how Darius, the head of the Persian Empire, decided to invade the Russian grasslands in 512 BC, then held by the Scythians. Scythians were a people whose way of life was probably similar to that of the Proto-Indo-Europeans, but further developed in that they had fully mastered the horse. They appear to have been milk drinkers early on: In fact, this is mentioned in the *Iliad*.[18]

When Darius invaded, the Scythians kept retreating farther and farther into the sea of grass: They had no cities or fields and thus had nothing to lose by retreating. Darius eventually realized that his expedition had been fruitless and turned back before his army ran out of supplies.[19]

Darius at least had a powerful state and a powerful army: He could cope with Scythian invasions, even if he couldn't conquer Scythia. Back in the early days of their expansion, the Indo-Europeans appear to have encountered farmers in the Balkans who had been farming since about 6000 BC, but who weren't under a powerful central government. Around 4200 BC, things went sour. Ancient village sites were abandoned, advanced work in metals and ceramics became rare, and the inhabitants shifted to easily defended sites such as caves, hilltops, and islands. We find an increasing number of Kurgan burials similar to those found earlier on the steppe. (Interestingly, the bodies in those Kurgan burials averaged almost four inches taller than the earlier peoples of the region—milk does a body good.)

We suspect that pre-state farmers had a lot of trouble with invading Indo-European pastoralists. It wasn't just that dairying was productive and conferred increased mobility. It made cattle very valuable, and cattle are far easier to steal than heaps of grain: They can walk. It looks as if the early Indo-Europeans spent a lot of time rustling each other's cattle, fighting over cattle, planning revenge for previous raids, and in general raising hell. They became a warrior society. That general tendency of pastoral society—a gift for causing trouble—was a key theme in

Eurasian history for millennia. The threat receded as agricultural peoples built strong states, intensified again in the Middle Ages as states weakened and steppe techniques improved (reaching an apogee with Genghis Khan), and ended only with the invention of gunpowder.

Our picture of the Indo-European expansion begins with a very rapid spread across the steppe as soon as the increased frequency of the lactase-persistence mutation became common enough to allow the switch to a dairying economy. This rapid spread would have resulted in a population that spoke similar dialects over a wide region all the way from the Ukraine to the Urals—similar because there hadn't been time for linguistic divergence. The wave of advance continued on into Europe, where dairying was ecologically competitive with early agriculture and produced a far more aggressive culture. Most likely, Indo-European culture also became more warlike as their mobility, superior numbers, and better nutrition allowed them to win battles more often than other peoples. Their victories, in turn, may have led to further advantages in military efficiency: Success feeds success.

Judging from their relatively low contribution to the European gene pool, Indo-Europeans appear to have practiced elite dominance, conquering rather than exterminating and replacing the previous inhabitants. A relatively small elite population can often impose its language on the rest of the population. In addition, the Indo-Europeans would have added the lactose-tolerance allele to the local mix. Although it appears to have been rare or nonexistent in Europe before the Indo-European invasions, it became common in those areas where a dairying economy was favored, particularly in northern Europe.[20] Indo-

European languages and culture spread past those regions in which dairying was favored—for example, into southern Europe and Iran—but strong states probably limited their expansion into the Middle East.

As much as anything, those peripheral expansions were probably driven by what might be called historical momentum: Peoples with a long record of success in war and raiding kept expanding even in areas where they had no special ecological advantages. Something similar happened when the Indo-Aryans moved into India: Internal weaknesses, possibly even collapse, of the Indus civilization may have allowed that expansion to occur. Today the LCT 13910-T lactase variant has reached almost 100 percent frequency in some parts of northern Europe; it is common in northern India and can even be found at low levels among some pastoral peoples of sub-Saharan Africa, such as the Fulani and Hausa.

Moreover, there is reason to think that this historical phenomenon has happened at least three times. Cattle herders of East Africa in the region of the Upper Nile and further south are lactase-tolerant milk drinkers due to a younger mutation of their own.[21] They, too, have expanded: They have become warlike, and there are fascinating parallels between their religions and social structure and those of the ancestral Indo-Europeans.[22] Another separate pair of mutations causing lactose tolerance happened in the Arabian peninsula, driven in this case by the domestication of camels. This may have been an important cause of the explosive growth of Islam and the Arab conquests of the seventh century AD and later.[23]

If this picture is correct, the occurrence of a single mutation in a particular group of pastoralists some 8,000 years ago eventually determined the spoken language of half of mankind. It may not be possible to reconcile this with Tolstoy's ideas of the unimportance of the individual in history. Of course, champions of individual importance have typically emphasized ideas, intelligence, and character—not digestion.

# 7

# MEDIEVAL EVOLUTION: HOW THE ASHKENAZI JEWS GOT THEIR SMARTS

The Ashkenazi Jews—the Jews of Europe—began as a distinct community about 1,200 years ago along the Rhine. The word "Ashkenaz" was the Hebrew name for Germany, so the Ashkenazim are literally "German Jews," although they later came to inhabit other areas, particularly Poland.

Today the Ashkenazi Jews, some 11 million strong, live throughout the world, with the largest concentrations in Israel and the United States. There are many other Jewish communities—such as the Sephardic Jews who once lived in Spain, the Mizrahi Jews of the Middle East and North Africa, and the Bene Israel of India—but the vast majority of the world's Jews are Ashkenazi.

They have had a surprisingly large influence on the world over the past couple of centuries, and they have played an outsized role in science, literature, and entertainment. Might they be smarter than other groups of people?

Apparently so. Ashkenazi Jews have the highest IQ of any ethnic group known. They average around 112–115, well above the European norm of 100. This fact has social significance, because IQ (as measured by IQ tests and their equivalents, like the Graduate Record Exam [GRE] or the Scholastic Aptitude Test [SAT]) is the best available predictor of success in academic subjects and many jobs.[1] Jews are just as successful in such jobs as their tested IQ would predict, and they are hugely overrepresented in those jobs and accomplishments with the highest cognitive demands.

We're not the first to notice this. Popular opinion has held that European Jews are smart for a long time. At the turn of the century in London, for example, Jews took a disproportionate share of prizes and awards in the school system.[2] This was not the case in classical times: Surviving writings from the ancient Greeks and Romans offer no hint that the Jews were considered unusually smart.

So why are the Ashkenazim especially intelligent?

To solve this puzzle, it may be useful to look at what we know about the DNA of the Ashkenazi Jews, because it turns out that they have another interesting characteristic. Namely, they have an unusual set of serious genetic diseases, such as Tay-Sachs disease, Gaucher's disease, familial dysautonomia, and two different forms of hereditary breast cancer (BRCA1 and BRCA2), and these diseases are up to 100 times more common in Ashkenazi Jews than in other European populations. For a long time, those disorders have posed another puzzle—why are they so common in this particular group?

We believe that these two puzzles have a single explanation. We propose that the Ashkenazi Jews have a genetic advantage in intelligence that arose from natural selection for success in white-collar occupations during their sojourn in northern Europe. Strong selection for intelligence also produced some unpleasant side effects, in the form of alleles that boost IQ in carriers while causing harm to homozygotes.

This kind of explanation is controversial, of course. It is true that many dismiss the idea that intelligence is measurable, is influenced by genes, or can vary from group to group. These criticisms and dismissals, interestingly, hardly ever come from scientists working in the area of cognitive testing and its outcomes: There is little or no controversy within the field. IQ tests work—they predict academic achievement and other life outcomes, and IQ scores are highly heritable. If genes influence intelligence, then, over time, a situation in which intelligence boosts fertility must result in higher intelligence. That simple logic is the very essence of the theory of evolution by natural selection: Genes that cause increased reproduction gradually become more and more common in a population.

## ASHKENAZI INTELLECTUAL PROMINENCE

Jewish intellectual prominence is striking. As we have said, Ashkenazi Jews are vastly overrepresented in science. Their numbers among prominent scientists are roughly ten times greater than you'd expect from their share of the population in the United States and Europe. Over the past two generations they have won more than a quarter of all Nobel science prizes, although they make up less than one-six-hundredth of the world's population. Although they represent less than 3 percent of the U.S. population, they won 27 percent of the U.S. Nobel Prizes in science during that period[3] and 25 percent of the A. M. Turing Awards (given annually by the Association for Computing Machinery).[4] Ashkenazi Jews account for half of twentieth-century world chess champions. American Jews are also overrepresented in other areas, such as business (where they account for about a fifth of CEOs[5]) and academia (where they make up about 22 percent of Ivy League students[6]). Although these statistics show intelligence in a broad range of disciplines, we emphasize measures of scientific and mathematical achievement in our present argument because we believe they are more objective measures than the others. Everyone agrees about what constitutes important discoveries in science and mathematics, whereas there are no comparable objective criteria to evaluate accomplishments in art and literature. Was Freudian theory, for example, a landmark achievement in psychology or the equivalent of the pet rock, a silly passing fad? We don't know (although we do have a strong suspicion), and we have no objective way of finding the answer.

The statistics about Ashkenazi accomplishment may seem pretty dry, but they're referring to people like Albert Einstein, who developed the special theory of relativity. This theory unified mechanics and electromagnetism and led to atomic energy. We're talking about John von Neumann, who was one of the developers of game theory and who played an important part in the Manhattan Project and in the development of the hydrogen bomb; and about Richard Feynman, Julian Schwinger, and

Murray Gell-Mann, who developed many of the most important ideas in particle physics.

The trend continues today, with scientists of Ashkenazi descent such as Ed Witten and Grigori Perelman. Witten, a professor of physics at the Institute for Advanced Study, has done important work in string theory and on emerging connections between mathematical physics and low-dimensional topology. He was the first physicist to win the Fields Medal (in 1990), the highest international award in mathematics, and won the Crafoord Prize, an international science award, in 2008. In 2002, Perelman, a Russian-Jewish mathematician, proved the Poincaré conjecture, the most famous unsolved problem in topology. For this work, he was offered—but refused—the Fields Medal (his refusal had to do with attempts by others to claim credit for his solution and his disappointment with the ethical standards of professional mathematics).

None of this means that typical Ashkenazi Jews are especially intelligent. Their average IQ is around 112, about three-quarters of a standard deviation above the European mean. However, a modest difference like this has a very strong impact on the number of individuals out at the far edge of the distribution because of the shape of the bell curve. It's enough to greatly increase the fraction of individuals with high intelligence.

This pattern among the Ashkenazim is far more interesting than most other kinds of human diversity. If a particular ethnic group had incredibly large ears, for example, we'd be amazed, but it wouldn't have much impact on our lives. Ideas originated by Ashkenazi Jews, such as special relativity and game theory, however, affect our lives every day, whether we know it or not. Their intelligence has influenced the world in important ways,

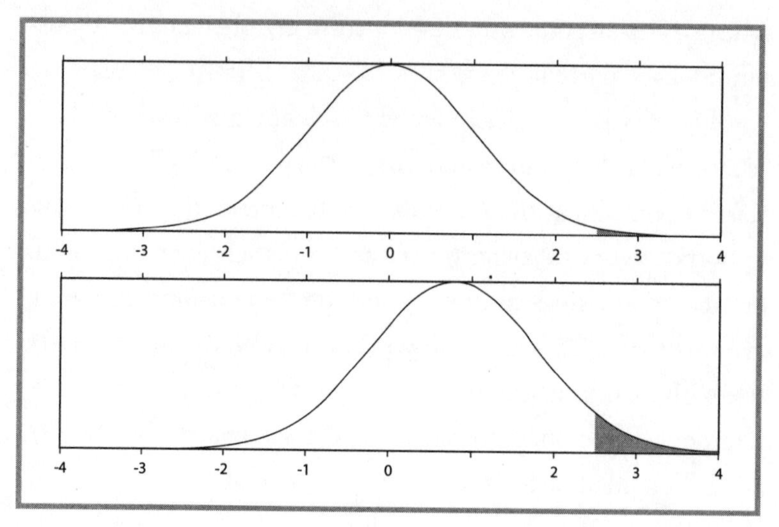


Two bell curves with different means

driving many of the most significant developments, advances, and creative works of our time.

Ashkenazi intellectual prominence is also very recent, in evolutionary terms. This high level of intellectual achievement among the Ashkenazi Jews is less than two centuries old.

## ASHKENAZI HISTORY UP TO 1800

The ancient Jewish population suffered remarkable vicissitudes—the Babylonian exile, the Hellenistic conquest and Hasmonean state, and the revolts against the Roman Empire, for example—but most of that history is irrelevant to our thesis, except to the extent that it helped create necessary cultural preconditions. That history is irrelevant because the Jews, in those days, were much like other people. Most Jews then were farmers, just like most other people in settled populations, and they must have experienced evolutionary pressures similar to those experienced by other agricultural peoples. They were not intellectually prominent at that time.

They made no contributions to the mathematics and protoscience of the classical era. A fair amount of classical commentary on the Jews has been preserved, and there is no sign that anyone then had the impression that Jews were unusually intelligent. By "no sign," we mean that there is apparently no single statement to that effect anywhere in preserved classical literature.[7] This is in strong contrast with the classical Greeks, whom everyone thought unusually clever.

The key cultural precondition among the Jews—key, that is, to later events among the Ashkenazim—was a pattern of social organization that required literacy, that strongly discouraged intermarriage, and that could propagate itself over long periods of time with little change. That pattern (Rabbinical Judaism) had not always existed but gradually emerged in the centuries after Titus's destruction of the Temple in the first revolt against the Roman Empire in AD 70. This happened first in Israel, then later in the Jewish community of Mesopotamia. It coincided with the development of the Talmud, a collection of writings about Jewish law, customs, and history. The Torah and the Talmud are the central documents of rabbinical Judaism.

Literacy, which does not itself require high intelligence, was probably important to the Jews in their shift from a nation to an urban occupational caste during and following the Diaspora, acting as an entrée to many urban professions in which they at first had no special biological advantages.[8] The prohibition against intermarriage mattered, because local selective pressures cannot change a population that freely mixes with neighbors. Intermarriage quickly dilutes the effect of beneficial alleles within a population, since the introduction of alleles from outside easily swamps the effects of selection within the group. Rabbinical Judaism's long-term stability was also key, since natural selection takes many generations to effect large changes.

In fact, pre-Diaspora Jewish genetics seems not to have been remarkable in any way. We make use of genetic markers indicating the amount of Middle Eastern ancestry among the Ashkenazim, but that is only important to our thesis insofar as it helps us estimate the extent of gene flow between the Ashkenazim and neighboring populations. In much the same way, the details of the Ashkenazi settlement of and migrations in Europe interest us because of their potential for creating genetic bottlenecks.

After the Bar-Kochba revolt of AD 135, most Jews lived outside of Israel. They were concentrated in the Parthian (later Sassanid) Empire and in the eastern half of the Roman Empire. There was a substantial population of Roman Jews, and there were other western settlements, such as Cologne, though these are poorly documented. The Jewish Diaspora in classical times was largely urban, but those Jews were on average poor; they were artisans and laborers rather than moneylenders or managers.[9] There is a temptation to project recent cultural patterns, such as Jewish concentration in finance or Talmudic scholarship, back into the past, well before those patterns came into existence—but this is a mistake. After the Muslim conquests, the majority of Jews lived under Islamic rule.

The Ashkenazim, the Jews living north of the Alps and the Pyrenees, appear in the historical record in the eighth and ninth centuries. Their origins are somewhat unclear. There are three different threads of history that may have led to the foundation of the Ashkenazi Jews in the centuries preceding, but the relative strengths of the theories are uncertain.

The first possibility is that the Ashkenazim—or some fraction of them—had already lived in France and the Rhineland for a long time, perhaps going back to Roman times. We know that there were Jews in Cologne around AD 300, and that there were Jews living in France under the Merovingian monarchs in the fifth and sixth centuries.[10] However, in 629 King Dagobert of the Franks ordered the Jews of his lands to convert, leave, or face execution. This conversion edict may have pushed them out of most of France. Certainly we hear little about French Jews for the next 150 years or so. The size and even the existence of this population is uncertain.

The second thread involves Jewish merchants originating from the lands of Islam as distant as Palestine and Iraq. The Carolingian kings encouraged and protected these merchants, who brought luxury items such as silks and spices from the East, according to Agobard of Lyons.[11] A few such traders served as interpreters on diplomatic missions; one brought Charlemagne an elephant from Haroun al-Rashid.

The third thread, generally thought to be the best supported, is that most of the founding Ashkenazi population migrated from southern Europe, especially Italy. There are accounts of particular Jews—both individuals and families—moving from Italy to this area in the early Middle Ages. One was the Kalonymus family, which is said to have migrated from Lucca in Italy to Mainz in 917.[12]

When they first appear in the historical record, the Ashkenazim are long-distance merchants who trade with the Muslim world. This is the beginning of a unique occupational pattern; there were no other European groups—or other Jewish groups, for that matter—who were noted for this. The majority of Jews had already given up agriculture, but the Jews of Islam, although urban, mostly worked in various crafts.[13] The Ashkenazim apparently seldom had such jobs. This pattern is detailed by one historian as follows: "Two entirely different patterns in the practice of crafts and their place in Jewish life and society are discernible throughout the Middle Ages. One characterizes the communities in countries around the Mediterranean, including in the south those in the continents of Asia and Africa, and in the north extending more or less to an imaginary demarcation line from the Pyrenees to the northern end of the Balkans. The other, in the Christian countries of Europe, was more or less north of the Pyrenees-Balkans line."[14] Furthermore, "North of the Pyrenees and in the Balkans crafts played a very small role as a Jewish occupation, from the inception of Jewish settlement there."

The Ashkenazi population, established in northern France by the early 900s, prospered and expanded. They settled in the Rhineland and then, after the Norman Conquest, in England. At first they were international merchants who acted as intermediaries with the Muslim world. As Muslims and Christians, especially Italians, increasingly found it possible to do business directly, Ashkenazi merchants moved more and more into local trade. When persecution became a serious problem and the security required for long-distance travel no longer existed, the Ashkenazim increasingly specialized in one occupation, finance, left open to them because of the Christian prohibition of usury. The majority of the Ashkenazim seem to have been moneylenders by 1100, and this pattern continued for several centuries.[15] Such occupations (trade and finance) had high IQ demands, and we know of no other population that had such a large fraction of cognitively demanding jobs for an extended period.

In some cases, we have fairly detailed records of Ashkenazi commercial activity. For example, concerning the Jews of Roussilon circa 1270: "The evidence is overwhelming that this rather substantial group of Jews supported itself by money lending, to the virtual exclusion of all other economic activities. Of the 228 adult male Jews mentioned in the registers, almost 80 percent appear as lenders to their Christian neighbors. Nor were loans by Jewish women (mostly widows) uncommon, and the capital of minors was often invested in a similar manner. Moreover, the Jews most active as moneylenders appear to have been the most respected members of the community."[16]

The Jews in this period were prosperous. Historian H. Ben-Sasson pointed out that "Western Europe suffered virtual famine for many years in the tenth and eleventh centuries, [but] there is no hint or echo of this in the Jewish sources of the region in this period. The city dweller lived at an aristocratic level, as befitted international merchants and honored local financiers."[17] Their standard of living was that of the lower nobility.[18] The Ashkenazi Jews were thus spared malnutrition and occasional famine. This helped Jewish populations recover from their losses due to persecution; it may have affected selective pressures as well.

And persecution was a very serious matter. Although the Jews of this region were prosperous, they were not safe. The first major crisis was the First Crusade of 1096, which resulted in the deaths of something like a quarter of the Jews in the Rhineland. Religious hostility, probably exacerbated by commercial rivalries, increased in Europe during this period, manifesting itself in the form of massacres and expulsions. This pattern of persecution kept the Ashkenazi Jews from overflowing their white-collar niche during the High Middle Ages, which otherwise would have happened fairly rapidly. It culminated in the expulsion of the Jews from most of Western Europe—from England in 1290, from France in 1394, and from various regions of Germany in the fifteenth century. The expulsions had greater effect on the demography of the region, in the long run, than massacres and persecutions. Jewish population growth rates were high due to both their prosperity and their belief system, as they favored large families, so their numbers tended to recover from attacks after a generation or two. But the potential for recovery decreased as Jews were excluded from more and more of Western Europe.

Many of the Jews who were expelled moved east, first to Austria, Bohemia, and Moravia, and later to the Polish-Lithuanian Commonwealth. The Polish rulers welcomed Jewish immigrants who could help modernize and reconstruct the country, which had been devastated by Mongol raids. Jews were welcomed as urban developers, investors, and initiators of trade. Other skilled immigrants were also welcomed, but some of those groups brought political risks, particularly the Germans because of their connection with the Teutonic Knights. The Jews were politically neutral and therefore safe.

As had been the case in Western Europe, the Jews of Poland had a very unusual occupational profile. None were farmers, and few were craftsmen, at least in the early centuries of that settlement. The very first to immigrate were mainly moneylenders, but that soon changed. They became tax-farmers (something like a freelance tax collector), toll-farmers, estate managers, and proprietors of mills and taverns. According to the historian B. D. Weinryb, in the middle of the fourteenth century "about 15 percent of the Jewish population were earners of wages, salaries and fees. The rest were independent owners of business enterprises."[19] They were the management class of the Polish-Lithuanian Commonwealth. Besides literacy, success in those specialized occupations depended upon skills similar to those of businessmen today, not least the ability to keep track of complex transactions and money flows.

Eventually, as the Ashkenazi population of the Polish-Lithuanian Commonwealth increased, more and more Jews became craftsmen—there are, after all, only so many managerial and financial slots. Still, for 800 to 900 years, from roughly 800 to 1650 or 1700, the great majority of the Ashkenazi Jews had managerial and financial jobs, jobs of high complexity, and were neither farmers nor craftsmen. In this they differed from all other settled peoples of which we have knowledge. In fact, it would have been impossible (back then) for the majority of any territorial ethnic group to have such white-collar jobs, because agricultural productivity would have been too low. Ninety percent of the population had to farm in order to produce enough to feed themselves and a thin crust of rulers, scribes, soldiers, craftsmen, and merchants. Selection for success at white-collar tasks could only have occurred if those scribes and merchants could somehow become an ethnic group, one defined by occupation rather than location.

Jews who were particularly good at these high-complexity jobs enjoyed increased reproductive success. As Weinryb noted: "More children survived to adulthood in affluent families than in less affluent ones. A number of genealogies of business leaders, prominent rabbis, community leaders, and the like—generally belonging to the more affluent classes—show that such people often had four, six, sometimes even eight or nine children who reached adulthood. . . . On the other hand, there are some indications that poorer families tended to be small ones. . . . It should also be added that overcrowding, which favors epidemics, was more prevalent among the poorer classes."[20] In short, Weinryb wrote, "the number of children surviving among Polish Jews seems to have varied considerably from one social level to another."[21] He also suggested that wealthier Jews were less crowded, as they lived in bigger houses; could keep their houses warmer; could afford wet-nurses; and had better access to rural refuges from epidemics. As an example, he cites a census of the town of Brody in 1764 showing that homeowner households had 1.2 children per adult member, while tenant households had 0.6.[22]

The occupational pattern of the Jews living in the Islamic world was different from that of the Ashkenazim. The Jews of Islam did not have a high concentration of white-collar occupations. Some had such jobs in some parts of the Islamic world, in some periods, but it seems it was never the case that most did in any given place and time. In part this was because other minority groups—Greek Christians, Armenians, and so on—competed successfully for these jobs, and in part it was because Muslims, valuing nonwarrior occupations more highly than medieval Christians did, took many of those jobs themselves. In addition, because there was less persecution overall, there were many *more* Jews in the Islamic world than in Europe—more Jews, in fact, than there were white-collar jobs.

In fact, to a large extent, and especially during years of relative Muslim decline from the fourteenth to the twentieth centuries, the Jews of Islam tended to have "dirty" jobs.[23] These included such tasks as cleaning cesspools and drying the contents for use as fuel—a common Jewish occupation in Morocco, Yemen, Iraq, Iran, and Central Asia. Jews were also found as tanners, butchers, and hangmen and in other disagreeable or despised occupations. Such jobs must have had low IQ elasticity: Brilliant tanners and hangmen almost certainly did not become rich.

## EMERGENCE OF INTELLECTUAL DIFFERENCES

The selection process we are interested in may have been under way during the Middle Ages, but its results were not yet evident in 1800. Of course, there were no IQ tests then, but there were as yet no Ashkenazi discoveries in science or mathematics either.

Perhaps if anti-Semitism had not prevented Jews from a wide range of career choices, this would not have been the case. Severe restrictions on Jewish occupations and participation in public life were just beginning to be lifted. But another reason they were not found among the early European scientists and mathematicians was that the Ashkenazim had been uninterested in natural philosophy for some centuries. Indeed, at times Ashkenazi leaders were positively hostile to such inquiries. Ever since the work of Maimonides in the twelfth century, the majority of Jewish religious leaders had tended to subordinate philosophy to the literal interpretation of the Torah (the first five books of the Hebrew Bible).

A *herem,* or religious ban, had been placed on Maimonides' philosophical work after his death in 1204, while Solomon ben

Abraham Adret, a rabbi and scholar at the beginning of the fourteenth century, issued another herem on "any member of the community who, being under twenty-five years, shall study the works of the Greeks on natural science and metaphysics."[24] Although the Ashkenazim put a tremendous amount of intellectual energy into Talmudic analysis during this period, we do not believe that the resulting body of work provides evidence of unusual mental acuity. Despite its historical significance and its importance to the Jewish culture and religion, it is fair to say that it hasn't drawn a great deal of interest outside of the Jewish community.

Another factor is that the Ashkenazi Jews were in the wrong place: Science and technology were sprouting in Western Europe, not in the Polish-Lithuanian Commonwealth. In fact, the Jews had been welcomed because Eastern Europe was backward and therefore in need of their skills.

All of these factors had begun to change by 1800. In 1791, France became the first country in Europe to grant legal equality to the Jews, and Napoleon's conquests spread that policy over much of Europe. Even in countries without full civil equality, such as Great Britain, Jews generally enjoyed increased rights. Movements within Judaism began to support Enlightenment values and escape from the ghetto, and things began to happen. As the British historian Eric Hobsbawm has said, "It is as though the lid had been removed from a pressure cooker."[25]

A trickle of Ashkenazi scientists and mathematicians began to appear in the first half of the nineteenth century—major talents like the eminent mathematicians Carl Jacobi and Leopold Kronecker. They originated in Germany, home to a comparatively small fraction of the Ashkenazim. Over time, more significant Jewish figures appeared in the arts and sciences. This might have happened more rapidly, except that the majority of Jews (after the partition of Poland) lived in Russia, which was slow to emancipate. Many Ashkenazim from Eastern Europe emigrated—some to Western Europe and some to the United States, Canada, Argentina, and South Africa. In those new lands they were much freer to exercise their talents, and by the beginning of the twentieth century, Ashkenazi Jews were playing a major role in all areas of science and mathematics. This trend strengthened in the second half of the twentieth century—even though the Nazis had murdered most of the Ashkenazim remaining in Europe—and it continues today.

## A GENETICALLY DISTINCT GROUP

One standard counterargument to any thesis suggesting that the Ashkenazi Jews are in some way biologically different or special is that they are adherents of a religion rather than a race or ethnic group in the strictest sense, and that therefore they cannot be genetically distinct. Some have brought up conversion as a mixing mechanism, often mentioning Elizabeth Taylor or Sammy Davis Jr., prominent converts to Judaism, as contemporary examples. Raphael and Jennifer Patai argued in *The Myth of the Jewish Race* that an inflow of genes from neighboring populations, via conversion, intermarriage, and illicit sex, kept Jewish populations from developing distinct genetic features.[26] It's true that Jews as a whole are not a single genetically distinct group; however, some subgroups are—in particular, the Ashkenazim. Strong evidence exists in the prevalence of genetic diseases like Tay-Sachs and others, but there is much more information these days as a result of new technologies for studying DNA. Take a look: SNPs don't lie.[27]

The plot shown on page 205 makes use of those alleles that are considerably more common in one group than in others to determine group membership. Ashkenazi Jews (represented by circles and squares, the cluster in the upper-right-hand corner) can easily be distinguished from the general European population (triangles). Irish, Scandinavians, Germans, and Brits occupy the upper left end of the archipelago, while Greeks and Italians are found at the lower left end. The Ashkenazi Jews are a distinct group; this is made especially clear by the cluster of dark squares, the group of Ashkenazi Jews with four Ashkenazi Jewish grandparents. For a very long time, Ashkenazi Jews (and most other Jewish groups as well) were endogamous, rarely marrying outside their faith or accepting converts. An endogamous group can remain genetically distinct, or become genetically different from neighboring peoples, if that social pattern persists. This is especially likely if some major fraction of the group's ancestors came from somewhere else (in this case, the Middle East) or if the selective pressures they've experienced have been different from those present in neighboring peoples.

There is reason to believe that a fair fraction (≈40 percent) of Ashkenazi ancestry is European, which we will discuss later, but it seems that for the most part those genes were added to the mix a long time ago, possibly back in the days of the Roman Empire. That notion is inherently plausible because many of the Jews in Rome arrived as enslaved prisoners of war, captured in the Great Revolt of AD 65–73 or in Bar-Kochba's revolt of AD 132–135. Many of those slaves eventually became freemen, and

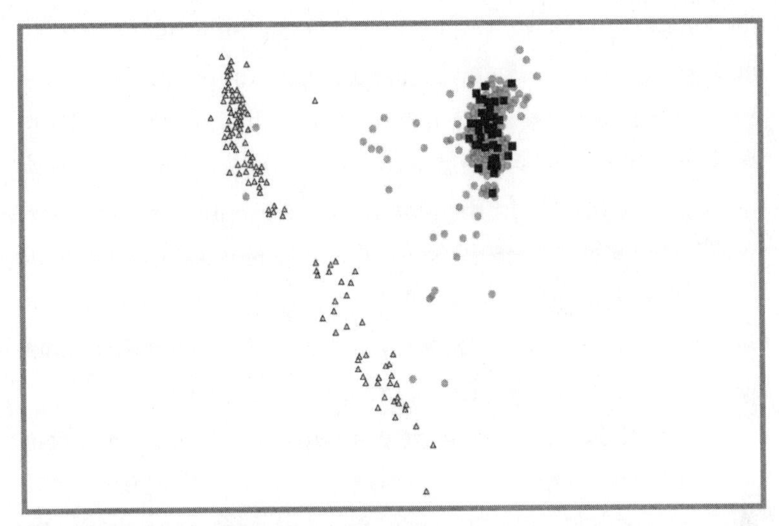

European genetic substructure analysis

it is likely that they were predominantly male. Many must have married local European women. There should therefore be a significant southern European component in the maternal ancestry of Roman Jews and, later, Ashkenazi Jews.

Admixture has not kept the Ashkenazim from becoming genetically distinct. Even if a population starts out as a mixture of two peoples, as in this case, becoming endogamous (ending intermarriage) and staying so for a long time ensures that the population will become homogeneous. If the population's ancestry is 60 percent Middle Eastern and 40 percent European, for example, a few dozen generations of endogamy will result in a population in which each individual's ancestry is quite close to 60 percent Middle Eastern and 40 percent European. In other words, you eventually get a population that has a flavor all its own—even more so if it experiences special selective pressures.

This means that if you look at the most informative parts of the genome, you can tell whether a certain individual is Ashkenazi (as opposed to, say, a non-Jewish Italian, Greek, or German) just about every time, particularly if all his or her recent ancestors are Jewish. In the plot, the circles represent Ashkenazi Jewish individuals, but the shaded circles represent individuals whose grandparents were all Ashkenazi Jews as well. That distinction matters, because Jews haven't been nearly as endogamous over the past century as they were during the Middle Ages.

Could these same methods distinguish the Ashkenazi from other Jewish groups, such as Moroccan Jews or Yemeni Jews? The answer is almost certainly yes. Although that particular measurement has not yet been made, it should be easy to make that distinction because the genetic distance between Ashkenazi Jews and Yemeni Jews is considerably larger than that between Ashkenazi Jews and Western Europeans.

Members of the general public sometimes believe that individual genetic profiles do not necessarily reflect nationality. Somebody who is Swedish, for example, might be genetically closer to someone from Japan than to another Swede, according to this view of things. If this was true, it would apply to a group like the Ashkenazi Jews as well, even though they are not quite a "nationality." However, that belief is false. In fact, a case where a person of one nationality is closer genetically to someone of a distant nationality than to his or her own compatriots *never* happens. If you're Swedish, *every* Swede (not counting recent immigrants) is genetically closer to you than any person in Japan.

It's possible to be closer to someone in Japan if you consider only one gene: Both of you might have the same blood type while your next-door neighbor might not, for example. Nevertheless, it is somewhat more likely that your neighbor will have the same blood type as you, since the frequency of blood types varies according to nationality. If you look over the whole genome—about 20,000 genes—with a match with your neighbor being somewhat more likely for every single gene, the chance of the overall match with someone Japanese being closer than the match with your neighbor becomes vanishingly small. Think of it this way: When you make a bet at the casino, the odds favor the house, although not overwhelmingly so. You might win that first bet—it's not that unlikely. But what is the chance that you'll win most of the time over the course of a year—win the majority out of thousands of bets, with the odds against you on every one? The probability of that happening is vanishingly small, which is why the house always wins in the long run.

As a practical matter, if you can distinguish between the members of two populations by looking at them, genetic analysis will be able to do so as well. And sometimes it will be able to make such distinctions when you *can't* tell by looking at them. The question as to whether the Ashkenazim are genetically distinct is now settled: We know from the data that they are. But that by itself it not enough to prove our thesis that they are more intelligent than the rest of us, or even that they are significantly different in any other way—not by a long shot. Being *measurably* different is not necessarily the same as being *significantly* different, and knowing that systematic genetic differences exist does not automatically tell us what their consequences are.

However, this initial set of data could have *disproved* our thesis. If the genetic evidence had indicated that the Ashkenazi

Jews could not constitute a genetically distinct group—if there had been significant continuing gene flow—then we would have to concede that our proposed mechanism (natural selection) could not have occurred. But the initial dataset did not disprove our thesis. So what does the genetic evidence say about Ashkenazi intelligence?

## NOTES ON IQ

IQ tests and scores are not in fact crucial to our thesis, but they are useful. Intellectual accomplishment is all that really matters: If people routinely won Nobel Prizes in physics with low IQ scores, or, for that matter, routinely aced calculus exams but flunked the IQ test, we'd junk the IQ tests. But that doesn't happen: IQ is an imperfect but useful measure of intelligence.

You'll frequently hear that we don't really know what intelligence is, that we don't know how to measure it, that IQ tests are biased, and that IQ scores don't predict anything, or that they don't predict anything outside of school. Often these complaints are salted with personal anecdotes about some acquaintance that had a high IQ score but was lazy, crazy, or suffered from unforgivable personal hygiene. And in recent years, other forms of intelligence have become all the rage. Daniel Goleman has written of "emotional intelligence" and "social intelligence," pointing out how they can help to predict job success and personal happiness. And other forms of intelligence have been proposed. In his 1993 book, Howard Gardner suggested that there are many types.[28] But the data hardly support these attempts to complexify cognitive testing. The supposed special kinds of intelligence don't predict anything useful or, when they do, predict only to the extent that they are correlated with general intelligence.

Yet IQ tests work in the sense that they predict performance. They were originally developed in order to predict how well children would do in school, and they do an excellent job of that. They also have moderate to high predictive power on many other questions, such as job performance, health, risk of accidental death, income, and other characteristics that may be less obvious, such as susceptibility to Alzheimer's disease. To make our position perfectly clear, we'd like to emphasize that saying IQ scores have *some* predictive power is not the same thing as saying that they determine everything.

Of course, exceptions don't make trends disappear. Muggsy Bogues may have played in the NBA while being 5 feet 3 inches tall, and there may be numerous individuals who are 6 feet 8 inches but were so clumsy on their high-school basketball teams that they sat on the bench the entire season. But in general, height still matters in basketball. It's not the only thing that matters, it doesn't absolutely determine success, but on average it makes a lot of difference. The same can be said of IQ: For most life events it's not as important as height is in basketball, but it's fairly important. Nor are IQ tests biased: They predict academic performance with the same accuracy in different ethnic groups.[29]

Moreover, IQ is highly heritable. What this means is that an individual's IQ is partially determined by genetic factors, so that it tends to be more similar to that of his or her parents and siblings than a randomly chosen person's IQ would be. Siblings with the same biological parents have similar IQs even when they are raised separately, whereas adopted siblings don't, even when raised together.

The same is true of height: Tall people tend to have taller-than-average children. In fact, IQ in adulthood is about as heritable as height. IQ in childhood, on the other hand, is less heritable and more susceptible to environmental influences. These effects of environment on the measured IQs of children, which disappear at or after puberty, are the basis for claims that IQ can be improved by interventions like Head Start.

Nongenetic factors also influence IQ, but for the most part, the ones that matter are not the ones people thought would matter. Prenatal care, breastfeeding, nutrition, access to early education, Mozart in the womb, and oat bran all have little or no effect. Surprisingly, the way in which a family raises children seems to have no effect on adult IQ. This argues against some popular environmental explanations for high intelligence among the Ashkenazi Jews—in particular, the notion that Jewish mothers have a special way of rearing children that boosts IQ.

## ASHKENAZI PSYCHOMETRICS

As noted earlier, Ashkenazi Jews have the highest average IQ of any ethnic group for which there are reliable data. Just how much higher is it? Many studies have found that it is 0.75 to 1.0 standard deviations above the general European average, corresponding to an IQ of 112–115.[30] Another, more recent study concluded that the advantage is slightly less, only half a standard deviation.[31] While the difference between the Ashkenazim and other northern Europeans in average IQ may not seem large, it leads to a large difference in the proportion of the two populations with very high IQs.[32] For example, if the mean northern European IQ is 100, the mean Ashkenazi IQ is 110, and the standard deviation in both populations is 15, then the number of northern Europeans with IQs greater than 140 should be 4 per 1,000, whereas 23 per 1,000 Ashkenazim should exceed the same threshold, about a sixfold difference. This is a general statistical effect, not something that happens only with IQ.

The fact that Ashkenazi Jews have high IQs on average and corresponding high academic ability has long been known. In 1900 in London, Jews took a disproportionate number of academic prizes and scholarships in spite of their poverty.[33] In the 1920s, a survey of IQ scores in three London schools with mixed Jewish and non-Jewish student bodies—one prosperous, one poor, and one very poor—showed that Jewish students, on average, had higher IQs than their schoolmates in each of the groups. The differences between Jews and non-Jews were all slightly less than one standard deviation, and the students at the poorest Jewish school in London had IQ scores equal to the overall city mean of non-Jewish children.[34]

That study, though conducted in 1928, is still important today because it contradicts a widely cited misrepresentation of a paper authored by researcher Henry Goddard in 1917.[35] Goddard gave IQ tests to people suspected of being retarded and found that the tests identified retarded Jews as well as retarded people of other groups. The psychologist Leon Kamin reported in 1974 that Goddard had found that Jews had low IQ scores. Kamin's reason for citing the study was to claim that Goddard and other IQ researchers of the 1920s had been biased against Jews and other minority groups. This erroneous analysis was picked up by many authors, including the well-known Harvard evolutionary biologist Stephen Jay Gould, who used it as evidence of the unreliability of IQ testing.[36] Gould seems to have believed that a popular impression that Jews had low IQs contributed to the passage of the Immigration Act of 1924, which was aimed at restricting immigration from southern and eastern Europe. However, by 1922 Jews already made up more than a fifth of Harvard undergraduates, and the Ivy League was already instituting admissions policies aimed at limiting Jewish admissions (the infamous "Jewish quotas"), which involved placing less emphasis on academic merit. If some people in the 1920s had the impression that Jews had low IQs, that impression cannot have been widely shared. The 1928 study of IQ in three London schools shows that in fact there were already researchers in the West who were noticing that Jews seemed to have higher IQs, on average, than the members of other groups.

But the achievements of Ashkenazi Jews are certainly not confined to IQ scores. They have an unusual ability profile when it comes to some other forms of testing as well. They have high verbal and mathematics scores on other types of standardized tests, though their visuospatial abilities—that is, their ability to rotate three-dimensional objects in their minds, for example—are typically somewhat lower, by about half a standard deviation, than the European average. The Ashkenazi pattern of success corresponds to this ability distribution—great success in mathematics and literature, more typical results in representational painting, sculpture, and architecture.

It is noteworthy that non-Ashkenazi Jews do not have high average IQ scores. Nor are they overrepresented in cognitively demanding fields like medicine, law, and academics. In Israel, Ashkenazi Jews, on average, score 14 points higher than Oriental Jews, almost a full standard deviation, which is 15 or 16 points on most IQ tests.[37] That difference means that the average non-Ashkenazi Jew in Israel would have an IQ score that would be at the 20th percentile among the Ashkenazim. Academic accomplishment in the two groups seems to vary in the same way, even among those born and raised in Israel: Third-generation Ashkenazi Jews in Israel are 2.5 to 3 times more likely to have graduated from college than third-generation Mizrahi Jews, for example (the ancestors of the Mizrahim moved to Israel from Asia and North Africa).[38]

## THE ASHKENAZI MUTATIONS

The best-known of the genetic diseases disproportionately affecting Ashkenazi Jews are Tay-Sachs disease, Gaucher's disease, and the breast-cancer mutations BRCA1 and BRCA2, but there are a number of others, such as Niemann-Pick disease, Canavan disease, and familial dysautonomia. Some of these cause neurological problems. And they're unusually common among Ashkenazi Jews—so common that they constitute an enduring puzzle in human genetics.

In principle, absent some special cause, genetic diseases like these should be rare. New mutations, some of which have bad effects, appear in every generation, but those that cause death or reduced fertility should be disappearing with every generation. Any particular harmful mutation should be rare; however, one in every twenty-five Ashkenazi Jews carries a copy of the Tay-Sachs mutation, which kills homozygotes in early childhood. This is an alarming rate.

The mutations that so frequently affect Ashkenazi Jews are mysterious in another way. Many of them fall into two categories or clusters involving particular metabolic pathways: They affect the same biological subsystem. Imagine a fat biochemistry textbook, where each page describes a different function or condition in human biochemistry: Most of the Ashkenazi diseases would be described on just two of those pages. The two most important genetic disease clusters among the Ashkenazim are the sphingolipid storage disorders (Tay-Sachs disease; Gaucher's disease; Niemann-Pick disease; and mucolipidosis, type IV) and the disorders of DNA repair (BRCA1 and BRCA2; Fanconi anemia, type C; and Bloom syndrome).

What is the explanation of this odd pattern? We know of only two mechanisms that can create high frequencies of dangerous, even lethal mutations: genetic drift in a bottleneck or natural selection.

## THE BOTTLENECK HYPOTHESIS

Most medical geneticists believe that these common Ashkenazi genetic diseases are a product of population bottlenecks. A population bottleneck occurs when a population goes through a period in which it is quite small. This often happens in the founding of a population. In a bottleneck, gene frequencies change almost randomly; just as you can get unrepresentative results (different from 50–50) when you flip a coin just a few times, or when you poll 20 people rather than 1,000, in a population bottleneck you get random changes that can affect large portions of the population in question.

When we say that a population is "small," we generally mean a few hundred individuals at most. Europe, for example, did not go through a bottleneck following the Black Death in the Middle Ages. The plague may have killed off half the population of Europe, but 40 million survivors is not a small number. It left plenty of genetic variation intact.

If a few lethal mutations became common in a bottleneck and a dramatic population expansion followed, we would see a large population with a surprising number of genetic diseases—diseases that were rare in most other populations. This has certainly happened in some cases. Amish communities that had very few original founders have seen this effect with their high incidence of several specific genetic disorders. It also happened in Pingelap, a Pacific island that was devastated by a typhoon around 1775, leaving about twenty survivors. Today almost 10 percent of the islanders suffer a form of severe color blindness.

Genetic diseases made common in a bottleneck are the product of chance, however, so there is no particular tendency for them to fall into a few metabolic pathways: They're scattered all over the biochemistry book, not concentrated on a few pages.

Our knowledge of human genetics has expanded rapidly over the past few years, and we now have good estimates of the total number of human genes (about 22,000) and the number of genes in different functional categories—in particular, the number involved in sphingolipid metabolism (108). We looked at twenty-one genetic diseases among the Ashkenazi and calculated the probability of finding four that affect sphingolipid metabolism, assuming randomness, in a given population. That probability was very low, less than 1 in 100,000. That can't be a coincidence.

We can say some other things about a population that has recently passed through a tight bottleneck. There would be overall genetic changes: reduced genetic variety in nuclear genes, increased genetic linkage, and increased genetic differences from other populations. All of these properties are measurable, and none have occurred among the Ashkenazi Jews.[39]

Finally, a genetic bottleneck would not increase a population's intelligence. If it was severe enough, it would almost certainly decrease intelligence as moderately deleterious genes became common.

Therefore, although bottlenecks can explain high frequencies of genetic disease in some cases, the bottleneck hypothesis cannot possibly explain the genetic data and the spectrum of genetic disease observed among the Ashkenazi Jews.

## NATURAL SELECTION

The alternative explanation is natural selection, the process by which some alleles cause increased reproduction in their bearers. Some gene variants have favorable effects in a given environment (in this case, the physical and social environment experienced by the Ashkenazi Jews during the Middle Ages), so that people with those variants have more children, on average, than others in that population. Those variants gradually become more common, ultimately leading to significant changes. In some cases, certain gene variants can have positive effects in individuals with one copy, and negative effects, such as disease, in individuals with two copies—the people with one copy have a "heterozygote advantage." As we have discussed earlier, the most famous example is sickle-cell anemia, where a mutation causing a very dangerous form of anemia in those carrying two copies has risen to high frequency in some parts of the world. There are a number of other malaria defenses of this sort that are expensive in terms of human health.

Clearly, natural selection can sustain quite high frequencies of serious, even lethal genetic diseases in some circumstances. Just as clearly, it can lead to a set of common mutations that cluster in a few metabolic pathways. This has happened with the malaria defenses: Many affect the hemoglobin molecule (sickle-cell, hemoglobin C, hemoglobin E, alpha- and beta-thalassemia), whereas others, such as G6PD deficiency or glycophorin C, affect other aspects of the red blood cell. Since the malaria parasite attacks red cells, this pattern is easy to understand.

Heterozygote advantage isn't confined to genetic defenses against malaria; it can also occur in other cases where certain traits are favored by selection. It seems that the key to such cases is that there has been strong selection (carriers have a big advantage) applied over a relatively short time period. Over longer periods, mutations with fewer side effects eventually occur and win out. The fact that heterozygote advantage can favor other traits is important, because we think that most of the characteristic Ashkenazi mutations are not defenses against infectious disease. One reason is that these mutations do not exist in neighboring populations—often literally people living across the street—that must have been exposed to very similar diseases. Instead, we think that the Ashkenazi mutations have something to do with Ashkenazi intelligence, and that they arose because of the unique natural-selection pressures the members of this group faced in their role as financiers in the European Middle Ages.

We see a clear example of heterozygote advantage in a trait other than disease resistance in whippets, a breed of dog similar to a small greyhound. Some whippets carry a mutated version of myostatin, a gene that limits muscle development. Dogs with this mutation grow more muscle. Whippets with one copy are faster, on average, than other whippets and often win races.[40] Those with two copies, called "bully whippets," are extremely muscular, have muscle spasms, and are not competitive as racers. One copy gives an advantage in a particular ability; two copies actually have a negative effect on the same ability.

Now, in order for natural selection to work in this way, the population has to be genetically isolated from its neighbors; otherwise it cannot become different. Admixture dilutes the effects of natural selection and can stop it in its tracks. You might compare it to boiling down soup while continually adding

Whippets with 0, 1, and 2 copies of the myostatin mutation

water—you won't get anywhere that way. As it happens, the Ashkenazi Jews *were* genetically isolated during the Middle Ages: not because of the Pacific Ocean, as happened with Pingelap, but owing to social reasons—internal rules against intermarriage combined with external prejudice.

For most of that period, both intermarriage with non-Jews and conversion to Judaism were very rare. That's certainly what the historical record indicates, but we can check that record using genetics. If we look at alleles that are clearly from the Middle East, we see that they account for a substantial fraction of Ashkenazi ancestry today: at least 50 percent, according to our analysis. This shows strong limits on the rate of genetic admixture, since even 2 percent mixing per generation over the past 2,000 years would have caused the Ashkenazim to become almost completely (80 percent) European. A continuing process of admixture (as opposed to a lot of admixture in the early stages) interferes the most with ongoing natural selection, but even if we make that pessimistic assumption, it looks as if the admixture rate was under 1 percent per generation—low enough to allow for the sort of natural selection we're suggesting. In fact, early European admixture might even have furthered the selective process, since even low levels of admixture can be an important source of beneficial alleles. More generally, the position of Israel at a natural crossroads, subject to invasion by Romans, Greeks, Persians, Babylonians, Assyrians, and Egyptians, may have resulted in unusually high genetic variety, which would also have furthered selection.

Already it's apparent that the Ashkenazi mutations were made common by natural selection, since there was sufficient genetic isolation for selection to occur, while nothing other than selection can explain the existence of common genetic diseases that are concentrated in a few metabolic pathways. You can also see why this kind of natural selection is usually confined to geographically isolated human populations, since such strict rules against intermarriage with neighboring groups are fairly unusual. But what kind of selection occurred? What traits were more valuable among the Ashkenazi than among their neighbors?

Well, we have some strong hints. What trait is accentuated among the Ashkenazim today? Are they as large as Samoans, as tall as the Tutsi, as milk-tolerant as the Dutch? No: Their special trait is intelligence.

The mutations themselves suggest this: Some of them *look* like IQ boosters, considering their effects on the development of the central nervous system. The sphingolipid mutations, in particular, have effects that could plausibly boost intelligence. In each, there is a buildup of some particular sphingolipid, a class of modified fat molecules that play a role in signal transmission and are especially common in neural tissues. Researchers have determined that elevated levels of those sphingolipids cause the growth of more connections among neurons, the basic cells of the central nervous system (see page 221).

There is a similar effect in Tay-Sachs disease: increased levels of a characteristic storage compound (GM2 ganglioside), which causes a marked increase in the growth of dendrites, the fine branches that connect neurons.[41] This increased dendritogenesis also occurs in Niemann-Pick type A cells and in animal models of Tay-Sachs disease and Niemann-Pick disease. These are the only known disease alleles that cause increased neural connections.

We also have evidence—not definitive—that some of the mutations common among the Ashkenazim may boost intelli-

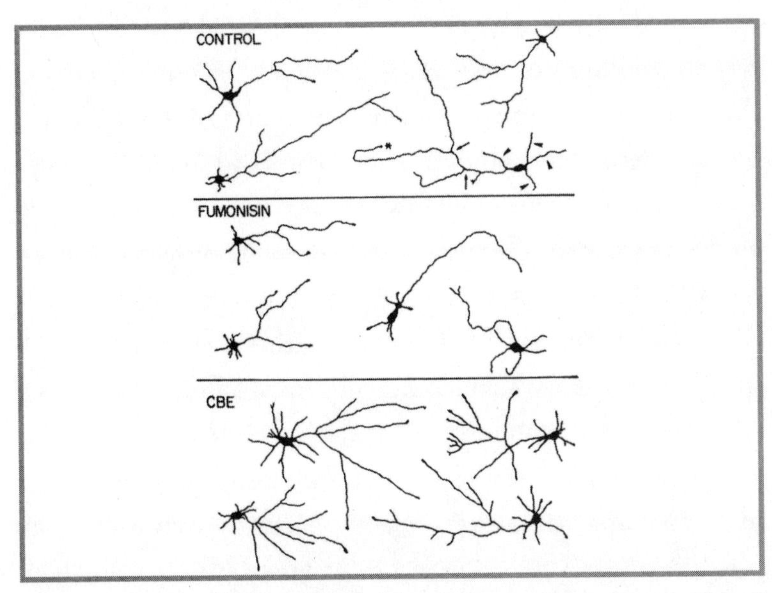


Axon growth in Gaucher's disease. The top frame shows normal cultured rat neurons, the middle frame shows stunted axonal growth resulting from reduced levels of glucosylceramide, and the bottom frame shows increased growth and branching in an experimental model of Gaucher's disease with increased levels of glucosylceramide. Glucosylceramide is the storage molecule involved in Gaucher's disease. Elevated levels promote growth and branching in axons, the transmission lines of the nervous system.

gence. We looked at the occupations of patients in Israel with Gaucher's disease, essentially all of whom were being treated at the Shaare Zedek Medical Centre in Jerusalem. These patients are much more likely to be engineers or scientists than the average Israeli Ashkenazi Jew—about eleven times more likely, in fact.[42] There are similar reports on torsion dystonia, another Ashkenazi genetic disease. Ever since it was first recognized, observers have commented on the unusual intelligence of the patients who suffer from it.

In 1976, Roswell Eldridge described early literature on torsion dystonia: One patient showed "an intellectual development far exceeding his age," and a second showed "extraordinary mental development for his age."[43] At least ten other reports in the literature have made similar comments. Eldridge studied fourteen Jewish torsion dystonia patients and found that their average IQ before the onset of symptoms was 121, compared to an average score of 111 in a control group of fourteen unrelated Jewish children matched for age, sex, and school district.[44] There are also reports of individuals with higher-than-average intelligence who have nonclassic congenital adrenal hyperplasia (CAH), another common Ashkenazi genetic disease. CAH, which causes increased exposure of the developing brain in a fetus to androgens (male sex hormones), is relatively mild compared to diseases like Tay-Sachs. At least seven studies show high IQ in CAH patients, parents, and siblings, ranging from 107 to 113. The gene frequency of CAH among the Ashkenazim is almost 20 percent.[45]

## HOW SELECTION HAPPENED

Our picture of how natural selection favored higher intelligence among European Jews in the Middle Ages relies upon three key observations. The first is that prosperous individuals had considerably more children, on average, than nonprosperous individuals in those days, as was then typical in most societies.[46] Second, Ashkenazi jobs were cognitively demanding, since the members of this group were essentially restricted to entrepreneurial and managerial roles as financiers, estate managers, tax farmers, and merchants. These are jobs that people with an IQ below 100 essentially cannot perform. Even low-level clerical jobs require an IQ of something like 90.[47] So, intelligence must have had greater rewards in those jobs than it does among farmers. This has to be true, really, since physical strength and endurance, which play a major part in success as a farmer, matter far less in finance and trade. If physical strength accounts for less of the variance, then cognitive and personality traits must account for more. Third, intelligence is significantly heritable. If the parents of the next generation are a little smarter than average, the next generation will be slightly smarter than the current one.

We can construct a scenario using IQ scores that illustrates this principle. We'll assume that parents of each generation averaged a single IQ point higher than the rest of the Ashkenazi adult population. In other words, let's suppose there was a modest tendency (mediated through economic success) for intelligent parents to have more surviving children than average parents—a tendency that certainly would not have been noticed at the time. If we assume a heritability of 30 percent for IQ, a very conservative assumption, then the average IQ of the Ashkenazi population would have increased by about a third of a point (0.30 point) per generation. Over forty generations, roughly 1,000 years, Ashkenazi IQ would have increased by 12 points. If we assume that the Ashkenazim began with a typical European IQ of 100 in the year AD 600, they would have reached an average IQ of 112 by 1600, just about what we see in the Ashkenazim today. This picture is consistent with observations of high verbal and mathematical scores among Ashkenazi Jews, paired with average or lower-than-average visuospatial scores. Verbal and mathematical talent would have helped medieval businessmen succeed, whereas visuospatial abilities were irrelevant.

There may well have been some selection for IQ among Europeans in general over this period. Christian merchants in

London or Rotterdam may have experienced selective pressures similar to those of the Ashkenazi Jews, but there was an important difference between those merchants and the Jewish population: Christian merchant families intermarried. The mixing would have caused extensive gene flow with the general population, the majority of whom were farmers. It seems that if IQ increased in the general European population, there was a greater increase among the Ashkenazim.

Our hypothesis also explains why certain things *didn't* happen—in particular, why we don't see high IQ scores and unusual intellectual achievement among other Jewish groups today. Although they, too, had very low rates of intermarriage, they never seem to have that high concentration of white-collar jobs that would have led to strong selection for verbal and mathematical intelligence. In part, this was because there were many more Jews in the Islamic world than in Christian Europe: With less persecution, there were more Jews than there were white-collar jobs. Our picture also explains why there's no real sign of unusually high intelligence among the Jews back in the days of the Roman Empire: The required events simply hadn't happened yet.

# CONCLUSION

Cultural innovation has been a driving force behind biological change in humans for a long time—certainly since the first use of tools some 2.5 million years ago. Natural selection acting on the hominid brain made those early innovations possible, and the innovations themselves led to further physical and mental changes.

Biological and cultural co-evolution was slow at first, at least by modern standards, but gradually things sped up. The archaeological record shows that our capacity for innovation continued to increase until, about 40,000 years ago, we were primed for what has been called the "human revolution" or the "creative explosion" of the Upper Paleolithic in Europe and northern

Asia. This sudden spurt in technology and art occurred shortly after modern humans expanded out of Africa—and it too must have involved biological changes, changes that we suspect were driven in part by genes stolen from the Neanderthals and other archaic humans, the previous occupants of Eurasia. Behavioral modernity led to even more change: Men made better tools and then, in turn, were reshaped by those tools over many generations.

With the development of agriculture, both cultural and biological evolution accelerated even further because that way of life made new demands on humans. Before agriculture humans had always been foragers: The huge population increase associated with agriculture resulted in more favorable mutations as well as more new ideas. This rapid evolution of our species following the spread of agriculture is indeed a 10,000 year explosion.

The explosion is ongoing: Human evolution didn't stop when anatomically modern humans appeared, or when they expanded out of Africa. It never stopped—and why would it? Evolutionary stasis requires a static environment, whereas behavioral modernity is all about innovation and change. Stability is exactly what we have not had. This should be obvious, but instead the human sciences have labored under the strange idea that evolution stopped 40,000 years ago.

All this means, has to mean, that biological change has been a key factor driving history. It has certainly not been the only factor, and it has been strangely intertwined with more traditional influences. Genetic changes like lactose tolerance have arisen and spread because of cultural innovations (such as the development of agriculture) as well as the random occurrence of the right mutations, and those genetic changes have in turn had their own cultural consequences. The expansion of the Indo-Europeans, the successful European settlement of the Americas and Australia, the failure of the "scramble for Africa," the entry of the Ashkenazi Jews onto the intellectual stage, possibly even the industrial revolution and the rise of science—all appear to be consequences of this endless dance between biological and cultural change.

If researchers in the human sciences continue to ignore the fact of ongoing natural selection, they will have thrown away the key to many important problems, turning puzzles into mysteries. Cortés, with 500 men, conquered and held an empire of millions. Try to explain this without invoking biological differences in disease resistance caused by ongoing evolution—it can't be done.

Thucydides in the fifth century BC said that human nature was unchanging and thus predictable, and many scientists today believe that human nature stopped changing tens of thousands of years ago. Historians seem to make the same assumption. In so doing, they're ignoring tremendous opportunities: not just in decoding the past, but in shaping the future as well. Continuing evolution over human history has been a vast natural experiment, an experiment that promises big payoffs in understanding, and then fighting, disease and mental illness. Limone sul Garda hid an important clue about human disease. With a million villages in the world, there must be many more such clues. Some of the results of history's experiments may even aid us in more ambitious efforts aimed at increasing human life spans and cognitive abilities.

It's time for researchers in the human sciences to shrug off the chains of dogmas like evolutionary stasis and "psychic unity." There's no time to lose—and there's a world to win.

# NOTES

## chapter one

1. Sargon of Akkad, about 4,000 years ago, built one of the first empires in what is now Iraq. Imhotep was an early Egyptian architect, engineer, and physician.

2. "Behavioral modernity" is anthropological slang for the cultural creativity thought to be characteristic of modern humans.

3. John Hawks, "Adaptive Evolution of Human Hearing and the Appearance of Language," 77th Annual Meeting of the American Association of Physical Anthropologists, April 11, 2008, Columbus, Ohio.

4. Stephen J. Gould, "The Spice of Life," *Leader to Leader* 15 (Winter 2000): 14–19.

5. Lyudmila N. Trut, "Early Canid Domestication: The Farm-Fox Experiment," *American Scientist* 161 (1999): 161.

6. Jack C. Schultz and Ted Floyd, "Desert Survivor," *Natural History* 108, no. 2 (1999): 24–29.

7. John Tooby and Leda Cosmides, "On the Universality of Human Nature and the Uniqueness of the Individual: The Role of Genetics and Adaptation," *Journal of Personality* 58, no. 1 (1990): 17–67.

8. John Tooby and Leda Cosmides, "Evolutionary Psychology: A Primer," University of California Santa Barbara, http://www.psych.ucsb.edu/research/cep/primer.html (accessed October 1, 2008).

9. M. Clifton, "Dog Attack Deaths and Maimings," 2008, from http://www.dogbitelaw.com/Dog%20Attacks%201982%20to%202006%20Clifton.pdf.

10. Yali Xue et al., "Spread of an Inactive Form of Caspase-12 in Humans Is Due to Recent Positive Selection," *American Journal of Human Genetics* 78, no. 4 (2006): 659–670.

11. George H. Perry et al., "Diet and the Evolution of Human Amylase Gene Copy Number Variation," *Nature Genetics* 39 (2007): 1256–1260.

12. Vincent Sarich is a professor of anthropology who played a key role in estimating the date of the divergence between humans and chimpanzees, Frank Miele is a senior editor at *Skeptic* magazine, and Chuck Lemme is a longtime reader of *Skeptic*.

13. Chuck Lemme, "Race and Sexual Selection," *Skeptic*, http://www.skeptic.com/eskeptic/05-03-22.html (accessed October 1, 2008).

14. Richard Lewontin, "The Apportionment of Human Diversity," *Evolutionary Biology* 6, no. 1 (1972): 381–398.

15. Mark Rieger, *Introduction to Fruit Crops* (New York: Food Products Press, 2006).

16. Koh-ichiro Yoshiura et al., "A SNP in the ABCC11 Gene Is the Determinant of Human Earwax Type," *Nature Genetics* 38 (2006): 324–330.

17. Laurent Keller and Kenneth G. Ross, "Selfish Genes: A Green Beard in the Red Fire Ant," *Nature* 394 (1998): 573; Michael J. B. Krieger and Kenneth G. Ross, "Identification of a Major Gene Regulating Complex Social Behavior," Science 295, no. 5553 (2002): 328–332.

18. John Hawks et al., "Recent Acceleration of Human Adaptive Evolution," *Proceedings of the National Academy of Sciences* 104, no. 52 (2007): 20753.

## chapter two

1. Roy C. Anderson, "The Ecological Relationships of Meningeal Worm and Native Cervids in North America," *Journal of Wildlife Diseases* 8, no. 4 (1972): 304–310.

2. Daniel M. Tompkins et al., "Parapoxvirus Causes a Deleterious Disease in Red Squirrels Associated with UK Population Declines," *Proceedings of the Royal Society, B: Biological Sciences* 269, no. 1490 (2002): 529.

3. Richard Klein, *The Human Career* (Chicago: University of Chicago Press, 1999), 524.

4. Richard Klein, *The Dawn of Human Culture* (New York: Wiley, 2002), 270.

5. Christopher S. Henshilwood et al., "Emergence of Modern Human Behavior: Middle Stone Age Engravings from South Africa," *Science* 295, no. 5558 (2002): 1278–1280.

6. Carleton S. Coon, *The Origin of Races* (New York: Alfred A. Knopf, 1963).

7. Chris Stringer and Peter Andrews, "Genetic and Fossil Evidence for the Origin of Modern Humans," *Science* 239, no. 4845 (1988): 1263–1268.

8. D. Curnoe et al., "Timing and Tempo of Primate Speciation," *Journal of Evolutionary Biology* 19, no. 1 (2006): 59–65.

9. Hilde Vervaecke and Linda Van Elsacker, "Hybrids between Common Chimpanzees (Pan troglodytes) and Pygmy Chimpanzees (Pan paniscus) in Captivity," *Mammalia* (Paris) 56, no. 4 (1992): 667–669.

10. Trenton W. Holliday, "Speciation by Distance and Temporal Overlap: A New Approach to Understanding Neanderthal Evolution," in *Neanderthals Revisited: New Approaches and Perspectives,* edited by T. Harrison and K. Harvati (New York: Sinauer, 2006).

11. J. Sweeney and T. Samansky, "Elements of Successful Facility Design: Marine Mammals," in *Conservation of Endangered Species in Captivity: An Interdisciplinary Approach,* edited by E. F. Gibbons, B. Durrant, and J. Demarest (New York: State University of New York Press, 1995).

12. On the supposed insignificance of Neanderthal admixture, see, for example, Chris Stringer and Peter Andrews, *The Complete World of Human Evolution* (London: Thames and Hudson, 2005), or Klein, *The Dawn of Human Culture.*

13. Milford Wolpoff et al., "Modern Human Ancestry at the Peripheries: A Test of the Replacement Theory," *Science* 291, no. 5502 (2001): 293–297.

14. It's conceivable that we already have: Martha Hamblin and Anna Di Rienzo of the University of Chicago found an extremely unusual version of a region near the Duffy gene in a few Italians, very different from that seen in other humans. M. T. Hamblin and A. Di Rienzo, "Detection of the Signature of Natural Selection in Humans: Evidence from the Duffy Blood Group Locus," *American Journal of Human Genetics* 66, no. 5 (2000): 1669–1679.

15. Amos Zayed and Charles W. Whitfield, "A Genome-Wide Signature of Positive Selection in Ancient and Recent Invasive Expansions of the Honey Bee Apis mellifera," *Proceedings of the National Academy of Sciences* 105, no. 9 (2008): 3421.

16. H. C. Stutz and L. K. Thomas, "Hybridization and Introgression in Cowania and Purshia," *Evolution* 18, no. 2 (1964): 183–195.

17. Brian Hare et al., "The Domestication of Social Cognition in Dogs," *Science* 298, no. 5598 (2002): 1634–1636.

18. Henry Harpending and Jay Sobus, "Sociopathy as an Adaptation," *Ethology and Sociobiology* 8, no. 3 (1987): 63–72.

19. Richard W. Bulliet, *The Camel and the Wheel* (Cambridge: Harvard University Press, 1975).

20. David W. Frayer, "Evolution at the European Edge: Neanderthal and Upper Paleolithic Relationships," *Préhistoire Européenne* 2 (1993): 9–69; David W. Frayer, "Perspectives on Neandertals as Ancestors," in *Conceptual Issues in Modern Human Origins Research,* edited by G. A. Clark and C. M. Willermet (New York: Aldine de Gruyter, 1998), 220–234.

21. Cidália Duarte et al., "The Early Upper Paleolithic Human Skeleton from the Abrigo do Lagar Velho (Portugal) and Modern Human Emergence in Iberia," *Proceedings of the National Academy of Sciences* 96 (1999): 7604–7609; Eric Trinkaus, "Early Modern Humans," *Annual Review of Anthropology* 34 (2005): 207–230.

22. D. Serre et al., "No Evidence of Neandertal mtDNA Contribution to Early Modern Humans," *PLoS Biology* 2, no. 3 (2004): 313–317; M. Currat and L. Excoffier, "Modern Humans Did Not Admix with Neanderthals during Their Range Expansion into Europe," *PLoS Biology* 2, no. 12 (2004): e21; T. D. Weaver and C. C. Roseman, "Ancient DNA, Late Neandertal Survival, and Modern-Human–Neandertal Genetic Admixture," *Current Anthropology* 46, no. 4 (2005): 677–683.

23. Daniel Garrigan et al., "Deep Haplotype Divergence and Long-Range Linkage Disequilibrium at Xp21.1 Provides Evidence That Humans Descend from a Structured Ancestral Population," *Genetics* 170 (2005a): 1849–1856; Daniel Garrigan et al., "Evidence for Archaic Asian Ancestry on the Human X Chromosome," *Molecular Biology and Evolution* 22 (2005b): 189–192; J. Hardy et al., "Evidence Suggesting That *Homo neanderthalensis* Contributed the H2 MAPT Haplotype to *Homo sapiens*," *Biochemical Society Transactions* 33 (2005): 582–585; Vincent Plagnol and Jeffrey D. Wall, "Possible Ancestral Structure in Human Populations," *PLoS Genetics* 2 (2006): e105; P. D. Evans et al., "Microcephalin, a Gene Regulating Brain Size, Continues to Evolve Adaptively in Humans," *Science* 309, no. 5741 (2005): 1717–1720; P. D. Evans et al., "Evidence That the Adaptive Allele of the Brain Size Gene Microcephalin Introgressed into *Homo sapiens* from an Archaic Homo Lineage," *Proceedings of the National Academy of Sciences* 103, no. 48 (2006): 18178.

24. Vincent Plagnol and Jeffrey D. Wall, "Possible Ancestral Structure in Human Populations," *PLoS Genetics* 2 (2006): e105.

25. Patrick D. Evans et al., "Evidence That the Adaptive Allele of the Brain Size Gene Microcephalin Introgressed into *Homo sapiens* from an Archaic Homo Lineage," *Proceedings of the National Academy of Sciences USA* 103, no. 48 (2006): 18178.

26. Graham K. Coop et al., "The Timing of Selection at the Human FOXP2 Gene," *Molecular Biology and Evolution* 25, no. 7 (2008): 1257.

27. Johannes Krause et al., "The Derived FOXP2 Variant of Modern Humans Was Shared with Neandertals," *Current Biology* 17, no. 21 (2007): 1908–1912.

28. James O'Connell and James Allen, "Dating the Colonization of Sahul (Pleistocene Australia—New Guinea): A Review of Recent Research," *Journal of Archaeological Science* 31, no. 6 (2004): 835–853.

## chapter three

1. John Hawks et al., "Recent Acceleration of Human Adaptive Evolution," *Proceedings of the National Academy of Sciences* 104, no. 52 (2007): 20753.

2. Jared Diamond, *Guns, Germs, and Steel: The Fates of Human Societies* (New York: W. W. Norton, 2005), 407.

3. Daniel Zohary and Maria Hopf, *Domestication of Plants in the Old World: The Origin and Spread of Cultivated Plants in West Asia, Europe, and the Nile Valley* (New York: Oxford University Press, 2001).

4. E. E. Thompson et al., "CYP3A Variation and the Evolution of Salt-Sensitivity Variants," *American Journal of Human Genetics* 75, no. 6 (2004): 1059–1069.

5. John Maynard Smith, *Evolution and the Theory of Games* (Cambridge: Cambridge University Press, 1982).

6. L. Cordain et al., "Original Communication: The Paradoxical Nature of Hunter-Gatherer Diets: Meat-Based, yet Non-Atherogenic," *European Journal of Clinical Nutrition* 56, no. 1 (2002): S42–S52.

7. Jared Diamond, "The Worst Mistake in the History of the Human Race," *Discover* 8, no. 5 (1987): 64–66.

8. Mark Nathan Cohen and George J. Armelagos, *Paleopathology at the Origins of Agriculture* (Academic Press, 1984); Clark Larsen, *Bioarchaeology: Interpreting Behavior from the Human Skeleton* (New York: Cambridge University Press, 1999).

9. Benjamin F. Voight et al., "A Map of Recent Positive Selection in the Human Genome," *PLoS Biology* 4, no. 3 (2006): e72; Eric T. Wang et al., "Global Landscape of Recent Inferred Darwinian Selection for *Homo sapiens*," *Proceedings of the National Academy of Sciences* 103, no. 1 (2006): 135–140.

10. A. Helgason et al., "Refining the Impact of TCF7L2 Gene Variants on Type 2 Diabetes and Adaptive Evolution," *Nature Genetics* 39 (2007): 218–225.

11. James Neel, "Diabetes Mellitus: A 'Thrifty' Genotype Rendered Detrimental by 'Progress'?" *American Journal of Human Genetics* 14 (1962): 353–362.

12. Lawrence H. Keeley, *War before Civilization: The Myth of the Peaceful Savage* (New York: Oxford University Press, 1996).

13. J. Burger et al., "Absence of the Lactase-Persistence-Associated Allele in Early Neolithic Europeans," *Proceedings of the National Academy of Sciences* 104, no. 10 (2007): 3736.

14. M. Fulge et al., "Lactose Persistence in Prehistoric Individuals," 8th International Conference on Ancient DNA and Associated Biomolecules, October 16–19, 2006, Lodz, Poland.

## chapter four

1. A. G. Maier et al., "*Plasmodium falciparum* Erythrocyte Invasion through Glycophorin C and Selection for Gerbich Negativity in Hman Populations," *Nature Medicine* 9, no. 1 (2003) 87–92.

2. T. J. Vullaimy et al., "Diverse Point Mutations in the Human Glucose-6-Phosphate Dehydrogenase Gene Cause Enzyme Deficiency and Mild or Severe Hemolytic Anemia," *Proceedings of the National Academy of Sciences* 85, no. 14 (1988): 5171–5175; Bernadette Modell and Matthew Darlison, "Global Epidemiology of Haemoglobin Disorders and Derived Service Indicators," *Bulletin of the World Health Organization* 86, no. 6 (2008): 480–487.

3. Richard Carter and Kamini N. Mendis, "Evolutionary and Historical Aspects of the Burden of Malaria," *Clinical Microbiology Reviews* 15 (2002): 564–594; Matthew A. Saunders et al., "The Extent of Linkage Disequilibrium Caused by Selection on G6PD in Humans," *Genetics* 171, no. 3 (2005): 1219–1229; Jun Ohashi et al., "Extended Linkage Disequilibrium surrounding the Hemoglobin E Variant due to Malarial Selection," *American Journal of Human Genetics* 74, no. 6 (2004): 1198–1208.

4. Meredith E. Protas et al., "Genetic Analysis of Cavefish Reveals Molecular Convergence in the Evolution of Albinism," *Nature Genetics* 38, no. 1 (2006): 107–111.

5. Gwynneth Stevens et al., "Oculocutaneous Albinism (OCA2) in Sub-Saharan Africa: Distribution of the Common 2.7-kb P Gene Deletion Mutation," *Human Genetics* 99, no. 4 (1997): 523–527.

6. Zanhua Yi et al., "A 122.5-Kilobase Deletion of the P Gene Underlies the High Prevalence of Oculocutaneous Albinism Type 2 in the Navajo Population," *American Journal of Human Genetics* 72, no. 1 (2003): 62–72.

7. John R. Baker, *Race* (New York: Oxford University Press, 1974), 279.

8. Benjamin F. Voight et al., "A Map of Recent Positive Selection in the Human Genome," *PLoS Biology* 4, no. 3 (2006): e72.

9. W. P. Rock et al., "A Cephalometric Comparison of Skulls from the Fourteenth, Sixteenth and Twentieth Centuries," *British Dental Journal* 200 (2006): 33–37.

10. Scott H. Williamson et al., "Localizing Recent Adaptive Evolution in the Human Genome," *PLoS Genetics* 10 (2007).

11. Ibid.

12. John Hawks, "Adaptive Evolution of Human Hearing and the Appearance of Language," Seventy-Seventh Annual Meeting of the American Association of Physical Anthropologists, April 11, 2008, Columbus, Ohio.

13. John Reader, *Africa: A Biography of the Continent* (New York: Knopf, 1998).

14. Gregory Clark, *A Farewell to Alms* (Princeton, N.J.: Princeton University Press, 2007).

15. J. Rousseau, *The Social Contract* (Baltimore: Penguin Classics, 1968).

16. John Robert McNeill and William H. McNeill, *The Human Web: A Bird's-Eye View of World History* (New York: W. W. Norton, 2003).

17. Robert C. Allen, "Agriculture and the Origins of the State in Ancient Egypt," *Explorations in Economic History* 34, no. 2 (1997): 135–154.

18. Laoise T. Moore et al., "A Y-Chromosome Signature of Hegemony in Gaelic Ireland," *American Journal of Human Genetic.* 78, no. 2 (2006): 334–338.

19. Rene Grousset, *The Empire of the Steppes: A History of Central Asia* (New Brunswick, N.J.: Rutgers University Press, 1970).

20. Mark G. Thomas et al., "Evidence for an Apartheid-Like Social Structure in Early Anglo-Saxon England," *Proceedings of the Royal Society B: Biological Sciences* 273, no. 1601 (2006): 2651–2657.

21. Robert Sallares et al., "The Spread of Malaria to Southern Europe in Antiquity: New Approaches to Old Problems," *Medical History* 48, no, 3 (2004):311–328.

22. The song "Is There for Honest Poverty," by Robert Burns, is more commonly known as "A Man's A Man for a' That," and is famous for its eloquent support of egalitarianism.

23. Jesse L. Byock, *Viking Age Iceland* (New York: Penguin, 2001).

24. Daniel G. MacArthur et al., "Loss of ACTN3 Gene Function Alters Mouse Muscle Metabolism and Shows Evidence of Positive Selection in Humans," *Nature Genetics* 39, no. 10 (2007): 1261.

25. William D. Hamilton, *Narrow Roads of Gene Land: The Collected Papers of W. D. Hamilton*, vol. 1, *Evolution of Social Behaviour* (New York: Oxford University Press, 1998).

26. Jared Diamond, *Guns, Germs, and Steel: The Fates of Human Societies* (New York: W. W. Norton, 2005), 417.

27. Valerie Bockstette et al., "States and Markets: The Advantage of an Early Start," *Journal of Economic Growth* 7 (2002): 347–369; Douglas O. Hibbs and Ola Olsson, "Geography, Biogeography, and Why Some Countries Are Rich and Others Are Poor," *Proceedings of the National Academy of Sciences* 101 (2004): 3715–3720.

28. Kay R. Jamison, *Touched with Fire: Manic-Depressive Illness and the Artistic Temperament* (New York: Free Press, 1993).

29. R. M. Nesse and George C. Williams, "Darwinian Medicine," *Life Science Research* 3 (1999): 1–17.

30. P. A. Hoodbhoy, "Science and the Islamic World—The Quest for Rapprochement," *Physics Today* 60, no. 8 (2007): 49–55.

## chapter five

1. Stephen L. Zegura et al., "High-Resolution SNPs and Microsatellite Haplotypes Point to a Single, Recent Entry of Native American Y Chromosomes into the Americas," *Molecular Biology and Evolution* 21, no. 1 (2004): 164–175.

2. Martin Richards, "The Neolithic Invasion of Europe," *Annual Review of Anthropology* 32, no. 1 (2003): 135–162; Isabelle Dupanloup et al., "Estimating the Impact of Prehistoric Admixture on the Genome of Europeans," *Molecular Biology and Evolution* 21, no. 7 (2004): 1361–1372.

3. Veronica L. Martinez-Marignac et al., "Admixture in Mexico City: Implications for Admixture Mapping of Type 2 Diabetes Genetic Risk Factors," *Human Genetics* 120, no. 6 (2007): 807–819.

4. Valter Gualandri et al., "AIMilano Apoprotein Identification of the Complete Kindred and Evidence of a Dominant Genetic Transmission," *American Journal of Human Genetics* 37, no. 6 (1985): 1083.

5. Cesare R. Sirtori et al., "Cardiovascular Status of Carriers of the Apolipoprotein A-IMilano Mutant," *Circulation* 103, no. 15 (2001): 1949–1954.

6. John K. Bielicki et al., "High Density Lipoprotein Particle Size Restriction in Apolipoprotein A-I (Milano) Transgenic Mice," *Journal of Lipid Research* 38, no. 11 (1997): 2314–2321.

7. Alan G. Fix, *Migration and Colonization in Human Microevolution* (New York: Cambridge University Press, 1999).

8. Georgi Hudjashov et al., "Revealing the Prehistoric Settlement of Australia by Y-Chromosome and mtDNA Analysis," *Proceedings of the National Academy of Sciences* 104, no. 21 (2007): 8726.

9. Jared Diamond, *The Third Chimpanzee: The Evolution and Future of the Human Animal* (New York: HarperCollins, 1992), 207.

10. Robert Sallares et al., "The Spread of Malaria to Southern Europe in Antiquity: New Approaches to Old Problems," *Medical History* 48, no. 3 (2004): 311–328.

11. Alessandro Achilli et al., "Mitochondrial DNA Variation of Modern Tuscans Supports the Near Eastern Origin of Etruscans," *American Journal of Human Genetics* 80, no 4 (2007): 759–768.

12. Marco Pellecchia et al., "The Mystery of Etruscan Origins: Novel Clues from *Bos Taurus* Mitochondrial DNA," *Proceedings of the Royal Society B: Biological Science* 274, no. 1614 (2007): 1175–1179.

13. Sadaf Firasat et al., "Y-chromosomal Evidence for a Limited Greek Contribution to the Pathan Population of Pakistan," *European Journal of Human Genetics* 15, no. 1 (2007): 121–126.

14. Tatiana Zerjal et al., "The Genetic Legacy of the Mongols," *American Journal of Human Genetics* 72, no. 3 (2003): 717–721.

15. Dio Cassius, *Roman History,* vol. 9, Books 71–80, translated by Earnest Cary and Herbert B. Foster, Loeb Classical Library, no. 177 (Cambridge: Harvard University Press, 1927).

16. Hans Eiberg et al., "Blue Eye Color in Humans May Be Caused by a Perfectly Associated Founder Mutation in a Regulatory Element Located within the HERC2 Gene Inhibiting OCA2 Expression," *Human Genetics* 123, no. 2 (2008): 177–187.

17. Edward Gibbon, *A History of the Decline and Fall of the Roman Empire*, vol. 3 (Philadelphia: B. F. French, 1830), 95.

## chapter six

1. Lawrence H. Keeley, *War before Civilization: The Myth of the Peaceful Savage* (New York: Oxford University Press, 1996).

2. L. Luca Cavalli-Sforza et al., *The History and Geography of Human Genes* (Princeton, N.J.: Princeton University Press, 1994).

3. Montgomery Slatkin and Christina A. Muirhead, "A Method for Estimating the Intensity of Overdominant Selection from the Distribution of Allele Frequencies," *Genetics* 156, no. 4 (2000): 2119–2126.

4. Kristin N. Harper et al., "On the Origin of the Treponematoses: A Phylogenetic Approach," *PLoS Neglected Tropical Diseases* 2, no. 1: e148 doi:10.1371/journal.pntd.0000148.

5. Noble David Cook, *Born to Die: Disease and New World Conquest, 1492–1650* (Cambridge: Cambridge University Press, 1998).

6. William H. McNeill, *Plagues and Peoples* (Garden City, N.Y.: Anchor Press/Doubleday, 1976).

7. Alison P. Galvani and Montgomery Slatkin, "Evaluating Plague and Smallpox as Historical Selective Pressures for the CCR5-delta32 HIV-Resistance Allele," *Proceedings of the National Academy of Sciences* 100, no. 25 (2003): 15276–15279.

8. Analabha Basu et al., "Genome-Wide Distribution of Ancestry in Mexican Americans," *Human Genetics,* DOI 10.1007/s00439-008-0541-5.

9. Ana Magdalena Hurtado et al., "The Epidemiology of Infectious Diseases among South American Indians: A Call for Guidelines for Ethical Research," *Current Anthropology* 42, no. 3 (2001): 425–432.

10. Richard Gordon, *Great Medical Disasters* (New York: Stein and Day, 1983), 41.

11. Henry Kamen, *Empire: How Spain Became a World Power, 1492–1763* (New York: HarperCollins, 2003), 205.

12. Charles R. Darwin, *The Voyage of the Beagle* (New York: Bantam Books, 1958), 376.

13. Aristotle, *History of Animals* (New York: Kessinger, 2004), 226.

14. Alfred W. Crosby, *Ecological Imperialism: The Biological Expansion of Europe, 900–1900* (New York: Cambridge University Press, 1986), 139.

15. James P. Mallory and Douglas Q. Adams, *The Oxford Introduction to Proto Indo European and the Proto Indo European World* (New York: Oxford University Press, 2006); David Anthony, *The Horse, the Wheel, and Language: How Bronze-Age Riders from the Eurasian Steppes Shaped the Modern World* (Princeton, N.J.: Princeton University Press, 2007).

16. James P. Mallory, *In Search of the Indo-Europeans: Language, Archaeology and Myth* (London: Thames and Hudson, 1989).

17. Raymond D. Crotty, *When Histories Collide: The Development and Impact of Individualistic Capitalism* (Walnut Creek, Calif.: AltaMira Press, 2001); Morton O. Cooper and W. J. Spillman, "Farmer's Bulletin No. 877—Human Food from an Acre of Staple Farm Products," *Farmers' Bulletin* of the U.S. Department of Agriculture (Washington, D.C.: Government Printing Office, 1919).

18. The line in the *Iliad* has been translated this way: "The Hippemolgi, whose diet is mares' milk . . ." Homer, translated by Stanley Lombardo, *Iliad* (Indianapolis: Hackett, 1997), 239.

19. Herodotus, translated by Aubrey de Selincourt, *The Histories* (New York: Penguin, 1972), 310–315.

20. J. Burger et al. "Absence of the Lactase-Persistence-Associated Allele in Early Neolithic Europeans," *Proceedings of the National Academy of Sciences* 104, no. 10 (2007): 3736.

21. Sarah A. Tishkoff et al., "Convergent Adaptation of Human Lactase Persistence in Africa and Europe," *Nature Genetics* 39, no. 1 (2007): 31–40.

22. Bruce Lincoln, *Priests, Warriors, and Cattle: A Study in the Ecology of Religions* (Berkeley: University of California Press, 1981).

23. N. S. Enattah et al., "Independent Introduction of Two Lactase-Persistence Alleles into Human Populations Reflects Different History of Adaptation to Milk Culture," *American Journal of Human Genetics* 82, no. 1 (2008): 57–72.

## chapter seven

1. Linda Gottfredson, "Logical Fallacies Used to Dismiss the Evidence on Intelligence Testing," in *The True Measure of Educational and*

*Psychological Tests: Correcting Fallacies about the Science of Testing,* edited by R. Phelps (Washington, D.C.: American Psychological Association, in press). Despite widespread condemnation of standardized tests, the data are unambiguous that they are the best available predictor of academic achievement and job success. There is essentially no controversy about this within the cognitive testing community.

2. Cyril Russell and Harry S. Lewis, *The Jew in London* (London: Harper Collins, 1900).

3. Data from the Nobel Foundation, http://nobelprize.org/ (accessed October 1, 2008).

4. "Jewish Recipients of the ACM A.M. Turing Award," http://www.jinfo.org/Computer_ACM_Turing.html (accessed October 1, 2008).

5. Richard S. Tedlow et al., "The American CEO in the Twentieth Century: Demography and Career Path," Harvard NOM Working Paper No. 03-21, Harvard Business School Working Paper No. 03-097, February 2003, available at SSRN: http://ssrn.com/abstract=383280 or DOI: 10.2139/ssrn.10.2139/ssrn.38328.

6. Data from Hillel International at http://www.hillel.org/HillelApps/JLOC/Search.aspx (accessed October 1, 2008).

7. Margaret H. Williams, *The Jews among the Greeks and Romans: A Diasporan Sourcebook* (Baltimore: Johns Hopkins University Press, 1998).

8. Maristella Botticini and Zvi Eckstein, "From Farmers to Merchants: A Human Capital Interpretation of Jewish Economic History," 2002, http://www.cepr.org/pubs/dps/DP3718.asp.

9. Mel Konner, *Unsettled: An Anthropology of the Jews* (New York: Viking Compass, 2003).

10. Williams, *The Jews among the Greeks and Romans;* Gregory of Tours, *The History of the Franks* (Harmondsworth, UK: Penguin, 1974).

11. Hayim Ben-Sasson, *A History of the Jewish People* (Cambridge: Harvard University Press, 1976).

12. Bernard D. Weinryb, *The Jews of Poland: A Social and Economic History of the Jewish Community in Poland from 1100 to 1800* (Philadelphia: Jewish Publication Society of America, 1973); Ben-Sasson, *A History of the Jewish People*; Zvi Ankori, "Origins and History of Ashkenazi Jewry (8th to 18th Century)," in *Genetic Diseases among Ashkenazi Jews,* edited by Richard M. Goodman and Arno G. Motulsky (New York: Raven Press, 1979); Eli Barnavi and Miriam Eliav-Feldon, *A Historical Atlas of the Jewish People: From the Time of the Patriarchs to the Present* (New York: Knopf, 1992).

13. Botticini and Eckstein, "From Farmers to Merchants."

14. Nachum Gross, *Economic History of the Jews* (New York: Schocken Books, 1975), 147, 150.

15. Marcus Arkin, *Aspects of Jewish Economic History* (Philadelphia: Jewish Publication Society of America, 1975); Ben-Sasson, *A History of the Jewish People.*

16. Arkin, *Aspects of Jewish Economic History*, 58.

17. Ben-Sasson, *A History of the Jewish People*, 401.

18. Norman Roth, *Medieval Jewish Civilization: An Encyclopedia.* Routledge Encyclopedias of the Middle Ages, vol. 7 (London: Routledge).

19. Weinryb, *The Jews of Poland*, 313.

20. Ibid.

21. Ibid., 115.

22. Ibid., 313.

23. Bernard Lewis, *The Jews of Islam* (Princeton, N.J.: Princeton University Press, 1984).

24. Jewish Encyclopedia, http://www.jewishencyclopedia.com/. The Jewish Encyclopedia was an encyclopedia originally published between 1901 and 1906 by Funk and Wagnalls. It contained over 15,000 articles in twelve volumes on the history and the state of Judaism and the Jews as of 1901. It is now a public domain resource.

25. Eric Hobsbawm, "Benefits of Diaspora," *London Review of Books*, October 20, 2005, http://www.lrb.co.uk/v27/n20/hobs01_.html.

26. Raphael Patai and Jennifer Patai, *The Myth of the Jewish Race* (New York: Scribner, 1975).

27. Tian et al., "Analysis and Application of European Genetic Substructure Using 300K SNP Information," *PLoS Genetics* 4, no. 1 (2008): e4. An SNP is a single nucleotide polymorphism. If we take two human chromosomes from a population and scan them side by side, there will be a single base pair difference approximately every 1,000 positions, on average. A site on the chromosome where two different bases are present in a population is called an SNP.

28. Howard Gardner, *Frames of Mind* (New York: Basic Books, 1993); Daniel Goleman, *Emotional Intelligence: Why It Can Matter More Than IQ* (New York: Bantam Books, 1995).

29. Arthur R. Jensen, *Bias in Mental Testing* (New York: Free Press, 1980).

30. Margaret E. Backman, "Patterns of Mental Abilities: Ethnic, Socioeconomic, and Sex Differences," *American Educational Research Journal* 9 (1972): 1–12; Boris Levinson, "A Comparison of the Performance of Monolingual and Bilingual Native-Born, Jewish Preschool Children of Traditional Parentage on Four Intelligence Tests," *Journal of Clinical Psychology* 15 (1959): 74–76; Julius S. Romanoff, "Birth Order, Family Size, and Sibling Spacing as Influences on Intelligence and Academic Abilities of Jewish Adolescents," Department of Psychology, Temple University, 1976.

31. Richard Lynn, "The Intelligence of American Jews," *Personality and Individual Differences* 26 (2004): 201–206.

32. James F. Crow, "Unequal by Nature: A Geneticist's Perspective on Human Differences," *Daedalus*, Winter 2002, 81–88.

33. Cyril Russell and Harry S. Lewis, *The Jew in London* (London: Harper Collins, 1900).

34. A. Hughes, "Jew and Gentiles: Their Intellectual and Temperamental Differences," *Eugenics Review* 18 (July 1928): 1–6.

35. Leon Kamin, *The Science and Politics of IQ* (Potomac, Md.: Erlbaum, 1974); H. Goddard, "Mental Tests and the Immigrant," *Journal of Delinquency* 2 (1917): 243–277.

36. Daniel Seligman, *A Question of Intelligence: The IQ Debate in America* (New York: Birch Lane Press, 1992).

37. Hanna David and Richard Lynn, "Intelligence Differences between European and Oriental Jews in Israel," *Journal of Biosocial Science* 39, no. 3 (2007): 465–473.

38. Yinon Cohen et al., "Ethnicity and Mixed Ethnicity: Educational Gaps among Israeli-Born Jews," *Ethnic and Racial Studies* 30, no. 5 (2007): 896–917.

39. Gregory Cochran et al., "Natural History of Ashkenazi Intelligence," *Journal of Biosocial Science* 38, no. 5 (2005): 659–693; A. B. Olshen et al., "Analysis of Genetic Variation in Ashkenazi Jews by High Density SNP Genotyping," *BMC Genetics* 9, no. 1 (2008): 14.

40. Dana S. Mosher et al., "A Mutation in the Myostatin Gene Increases Muscle Mass and Enhances Racing Performance in Heterozygote Dogs," *PLoS Genetics* 3, no. 5 (2007): e79.

41. Steven U. Walkley, "Neurobiology and Cellular Pathogenesis of Glycolipid Storage Diseases," *Philosophical Transactions of the Royal Society London B* 358 (2003): 893–904; Steven U. Walkley et al., "Gangliosides as Modulators of Dendritogenesis in Normal and Storage Disease-Affected Pyramidal Neurons," *Cerebral Cortex* 10 (2000): 1028–1037.

42. Cochran et al., "Natural History of Ashkenazi Intelligence."

43. Roswell Eldridge, "Edward Flatau, Wladyslaw Sterling, Torsion Spasm in Jewish Children, and the Early History of Human Genetics," *Advances in Neurology* 14 (1976): 105–114.

44. Roswell Eldridge, "Torsion Dystonias: Genetic and Clinical Studies," *Neurology* 11 (1970): 1–78; Eldridge, "Edward Flatau, Wladyslaw Sterling, Torsion Spasm."

45. Maria I. New and R. C. Wilson, "Steroid Disorders in Children: Congenital Adrenal Hyperplasia and Apparent Mineralocorticoid Excess," *Proceedings of the National Academy of Sciences of the USA* 96 (1999): 12790–12797.

46. At the beginning of the nineteenth century in Europe the old relationship between wealth and number of offspring began to reverse, and it remains reversed today in industrial societies. This accompanied a reduction in death rates and, somewhat later, birth rates among all wealth classes, called the "demographic transition."

47. Linda S. Gottfredson, "Why *g* Matters: The Complexity of Everyday Life," *Intelligence* 24 (1997): 79–132.

# GLOSSARY

**Adaptive:** An adjective that is widely used in evolutionary biology but that is nowhere precisely defined. In general, a trait is called adaptive if it increases the fitness of its bearers: Thus, white skin may be adaptive in cold, northern climates where there is limited opportunity for exposure to sunlight and hence vitamin D synthesis.

**Akkadian Empire:** A polity that was prominent in Iraq two thousand years before the Common Era. The empire had a system of roads and a regular postal service, and there was intense use of clay seals as postage stamps.

**Alans:** Horse nomads, a subgroup of Sarmatians, who accompanied the Vandals in their invasion of the Roman Empire and later wanderings in later Roman times and during the dark ages. *See also* Sarmatians; Vandals.

**Allele:** An alternate form of a gene; also, a particular sequence of nucleotides occupying a given position on a chromosome. The position on a chromosome is called a *locus* (pl. *loci*). Different sequences that occur at the same locus in the population are called *allelic* to each other or simply *alleles*. Thus, A, B, and O are alleles at the ABO blood-group locus.

**Amino acid:** The basic building blocks of proteins. There are twenty standard amino acids—twenty-one if you count selenocysteine.

**Anatomically modern humans (AMH):** Creatures who looked much like people today, who appeared in northeastern Africa up to 200,000 years ago. From then until about 45,000 years ago, there are occasional traces of innovation in the archaeological record—beads, ochre, or an occasional new high-quality tool, but little happens until about 45,000 years ago when our human ancestors left Africa to colonize Europe and Asia north of the Himalayas, in one branch, and Australia, parts of Indonesia, and near Pacific islands in another branch. Some would say that the 200,000-year-old Africans were the first anatomically modern humans; others would say that the colonizers of 45,000 years ago were.

**Archaic humans:** Precursors of anatomically modern humans in Africa, Europe, and Asia, including the Neanderthals of Europe. *Homo* *erectus* is the name usually given to the human ancestor of 1.8 million years ago until about 300,000 years ago. After that, *H. erectus* began to exhibit several important changes: The first prepared-core toolmaking traditions appeared, along with hearths for fire, and the brain became larger, until it was essentially the same size as or slightly larger than that of humans today. It is these apparently larger-brained versions of *Homo erectus* that are called "archaic humans."

**Assyrian Empire:** A Bronze Age empire centered in northern Mesopotamia, north of Babylon. The Assyrians spoke a Semitic language.

**Atlatl:** A spear-throwing device with a handle on one end along with a spur or cup that the spear rests against. An atlatl can cast a dart more than 100 yards. The Australian version is known as a *woomera.*

**Aurignacian:** The earliest of several cultural traditions in Europe associated with the anatomically modern human invaders from Africa. Aurignacian culture had early cave art, sculpture, and fine stone and bone tools.

**Autosome:** A chromosome other than the sex chromosomes. Twenty-two of the twenty-three pairs of chromosomes in human DNA are autosomal; two copies of each autosomal chromosome are found in both sexes. The chromosomes of the other pair, which can be either type X or Y, are called sex chromosomes.

**Axon:** A long fiber or projection of a nerve cell that carries nerve impulses away from the cell body.

**Behavioral modernity:** The set of practices and cultural elaborations characterizing modern humans, as opposed to archaic humans. There is an ongoing effort in some disciplines to draw a line between archaic and modern humans, but different theorists place that line at different points in the history of our species. Some would draw it several hundred thousand years ago in Africa when creatures who look like us appeared on the scene; others would draw the line around 45,000 years ago, when a lot of art, decoration, clothing, sculpture, and new technology appeared rather suddenly, especially among populations that left Africa.

**Berbers:** Aboriginal inhabitants of Africa north of the Sahara. The Berbers look much more like Europeans than sub-Saharan Africans do.

**Blade:** A stone tool made from a flake that is more than twice as long as it is wide. The toolkit of the anatomically modern invaders of Europe about 40,000 years ago had many more blades than the Neanderthals' toolkit did.

**Bottleneck:** A severe restriction in population size that leads to a reduction in genetic diversity. The effect of a bottleneck depends on the size of the population when it was small and the length of time that it was small. In general, in human history a reduction to a total population size in the hundreds would leave a genetic trace; a reduction to a population size in the thousands would not be a significant bottleneck unless it persisted for tens of thousands of years.

**Carbon-14 dating:** A method of dating paleontological and archaeological specimens by measuring the ratio of stable to unstable carbon isotopes present in the specimen. Carbon in the atmosphere contains an unstable isotope that breaks down at a constant rate and is generated by solar activity. When a living organism dies, there is no longer carbon exchange with the atmosphere, and the fraction of the unstable isotope decreases with time. By measuring the ratio of stable to unstable carbon, one can estimate the time of death. The method is considered accurate for events from several centuries to about 45,000 years in the past. *See also* Isotope.

**Carrying capacity:** The maximum population that can be sustained over the long run. With humans, this depends upon which skills and tools are known. The same piece of land would have a lower carrying capacity for foragers than for farmers.

**Centromere:** A central region in the chromosome involved in cell division. It consists of largely repetitive DNA and has few genes.

**Châtelperronian:** A stone-tool tradition found in Europe after the invasion of anatomically modern humans from Africa. It shows some similarities to technologies of the Neanderthals and has been associated with Neanderthal remains. Many believe that it represents a Neanderthal imitation of the technology of the invaders.

**Chromosome:** A very long DNA molecule, along with associated protective proteins. Humans have forty-six chromosomes in twenty-three pairs, one of each pair from the mother, one from the father. The last pair, the sex chromosomes, would be a pair of two X's in females or an X and a Y chromosome in males.

**Codon:** A sequence of three nucleotides. These three-base sequences designate particular amino acids (most of them—sixty out of sixty-four) or initiation or cessation of protein assembly.

**Dendrite:** Short, highly branched extensions of a neuron. Dendrites form synaptic contacts with other neurons.

**Diploid:** Organisms whose cells carry two copies of the genetic blueprint, ordinarily one from the mother and one from the father. Diploidy is associated with sexual reproduction.

**DNA:** Deoxyribonucleic acid, a nucleic acid molecule that contains the genetic recipe used in the growth and function of living organisms.

**Dominant:** The phenotypic effect of an allele at a locus when there is only a single copy. For example, people with blood group A have either two

A alleles, or else an A and an O; in the latter case, A is called dominant to O.

**Eemian interglacial:** The interglacial era before the current one (Holocene). It began about 131,000 years ago, and temperatures had returned by 114,000 years ago. At its peak, temperatures in the Northern Hemisphere were 1–2°C warmer than today.

**Etruscans:** An ancient people of Italy, concentrated in the area north of Rome now known as Tuscany. They had, at minimum, an important influence on the development of Rome, and they may have founded the city. They spoke an undeciphered non-Indo-European language, and historical and genetic evidence suggests that they originated in Anatolia.

**Exponential growth:** Growth that is proportional to the amount present, like compound interest. Interest that is continuously compounded yields an exponential growth of money. Under exponential growth, the doubling time is constant. So, for example, it takes the same amount of time to go from 100 to 200 as it takes to go from 1,000 to 2,000.

**Fitness:** The rate of reproduction of an entity. The fitness of an individual is the genetic contribution of that individual to the next generation. The fitness of a gene, or a chunk of DNA, is the rate of reproduction of that gene.

**Fixation:** The state in which all copies of a given gene in a population are identical.

**Gaucher's disease:** A lysosomal storage disease that is unusually common (100 times the norm) in Ashkenazi Jews. Homozygotes experience an illness of varying severity, mostly due to the accumulation of the fatty substance glucosylceramide in tissues.

**Gene:** A string of nucleic acids that does something biologically useful. Often the useful product is messenger RNA that codes for a protein, but some genes produce structural RNA or regulate expression of other genes.

**Genotype:** An individual's genetic pattern, as opposed to his phenotype, which is his visible characteristic. For example, someone with the phenotype of blood group A could have either the AA or the AO genotype.

**Group selection:** Selection favoring well-adapted groups rather than well-adapted individuals. The classic formulation is this: Group A is full of altruists, and group B is full of selfish individuals. Group A grows faster, but since the groups belong to the same species, selfish individuals from group B always infiltrate group A and take advantage of the altruism there. In the end, selfish individuals predominate. There is a broad consensus that group selection is not an important force in evolution. This means that evolution does not lead to individuals who are altruistic or who do things "for the good of the species."

**Haploid:** An organism that carries a single copy of the genetic code, that is, its blueprint. Most complex animals and plants are instead diploid. *See also* Diploid.

**Haplotype:** A set of single nucleotide polymorphisms (SNPs) near each other on a chromosome that are statistically associated. *See also* SNP.

**HapMap:** A genetic dataset that records the common single-nucleotide variants in a number of individuals drawn from Europe, sub-Saharan Africa, and East Asia.

**Heritability:** The proportion of variation of a trait that is caused by genetic variation. Note that this is not fixed for any trait, since it depends on the amount of genetic variation and the diversity of environmental or other effects. Skin color is more heritable in New York than it is in Stockholm because the genetic variation is greater in New York.

**Heterozygote:** An individual with two different alleles of a given gene at a locus.

**Holocene:** The interglacial period we are currently in. It started about 11,500 years ago.

***Homo erectus*:** A human ancestor that left Africa over 1.5 million years ago to occupy much of the temperate and tropical Old World. Below the neck they resembled extremely rugged modern humans. Their brains were approximately two-thirds the size of ours and their skulls were thick and heavy, with prominent bony architecture around the eyes and heavy jaws and teeth.

***Homo heidelbergensis*:** An early European version of *Homo erectus*. Whether or not to give the European forms this separate name is a matter of taste and convenience.

***Homo neanderthalensis*:** *See* Neanderthal.

***Homo sapiens*:** The proper name of our species, anatomically modern humans.

**Homozygote:** An individual with two identical copies of a given gene at a locus.

**Indo-European:** A family of related languages. Most European languages are Indo-European, as are Persian and the languages of northern India. The Indo-European range once extended into western China. There are several plausible theories of Indo-European origins, the favorite being that the original Indo-European speakers (called Proto-Indo-Europeans) were mounted agro-pastoralist invaders from the steppes of the Ukraine.

**Indus civilization:** An ancient civilization that flourished in Pakistan and western India from 2600 to 1900 BC. Cuneiform records and archaeological finds indicate extensive maritime trade between this civilization and ancient Mesopotamia.

**Introgression:** The movement of a gene or genes from one species to another.

**Isotope:** One of several forms of an element that differ in weight. Isotopes have the same number of protons but differing numbers of neutrons. *See also* Carbon-14 dating.

**Lactase:** The digestive enzyme that breaks down lactose. People who lack lactase often suffer gastrointestinal discomfort or worse when they drink fresh milk.

**Lactose:** A sugar found in milk.

**Linkage disequilibrium:** Statistical association in a population between nearby genetic variants, ordinarily SNPs. Chromosomes break and recombine over time, so that diversity along a chromosome becomes randomized in a population. Ordinarily, for example, the pattern of SNPs at a chromosomal location predicts little about the pattern 100,000 base positions away. Various evolutionary forces, however, can lead to a correlation over large chromosomal distances, and this is called linkage disequilibrium.

**Locus:** A particular place on a chromosome. Since humans are diploids, we have two copies of the genetic material at every locus.

**Loss of function:** Impairment in the functioning of a gene due to a mutation. A mutation of a gene that breaks it or impairs its function is called a *loss-of-function mutation.* A familiar example is any of several genes that cause light skin color in Europeans. These are essentially broken African genes. We see many loss-of-function mutations in genetics because it is easy to break a gene.

**Malthusian trap:** A situation in which diminishing returns prevent gains in average welfare in a population because any improvements in technology or production are offset by population increase.

**Mitochondrial DNA:** Genetic material in organelles in the cell called mitochondria. Mitochondrial DNA (mtDNA) is transmitted in the cytoplasm outside the nucleus; hence it is only transmitted through females. Since any individual has only one inherited copy of mtDNA, the mtDNA system is called a haploid (as opposed to diploid) system. *See also* Diploid; Haploid.

**Mousterian:** The culture and tool tradition associated with Neanderthals in Europe and near Asia. It is characterized by sophisticated flake tools made from prepared cores, but it also retains earlier technologies.

**Mutation:** A change in genetic material. Mutation is the ultimate source of all genetic diversity.

**Natural selection:** The process by which the frequency of alleles in a population is changed by factors other than sampling error. Usually this means that changes in the frequency of an allele are driven by its effects on individual fitness.

**Neanderthal:** Archaic humans who occupied western Eurasia from several hundred thousand years ago to 35,000 years ago when they went extinct, presumably in competition with anatomically modern humans.

**Neolithic:** Literally "New Stone Age"; the period beginning with the advent of farming and ending when metal tools became widely used. In the Middle East, it began about 9000 BC and ended about 4500 BC. *See also* Paleolithic.

**Neuron:** A nerve cell.

**Nucleotide:** Chemical compounds that are the structural units of DNA and RNA. They're made up of a unit called a *base* that is linked to a sugar and one or more phosphate groups. The four bases found in DNA are adenine, cytosine, guanine, and thymine. A chromosome is made of millions of linked nucleotides.

**Paleolithic:** Literally, "Old Stone Age"; the period beginning over 2 million years ago and represented by the earliest stone-tool traditions from the Oldowan tradition and including the Upper Paleolithic tools made by anatomically modern humans as recently as 10,000 years ago in Europe. The period ended with the advent of farming in around 9000 BC. *See also* Neolithic.

**Parsis:** The Zoroastrian community in India, descended (in part) from Persians who left Iran after the Arab conquest.

**Pathans:** The largest ethnic group in Afghanistan and the second-largest in Pakistan. Pathans are Muslims, speak an east Iranian language, and follow an elaborate honor code known as Pashtunwali. They are also called Afghans and Pushtuns.

**Pleistocene:** A geological epoch that began about 1.8 million years ago and, conventionally, ended with the end of the last Ice Age about 12,000 years ago.

**Positive selection:** Advantage that occurs when natural selection favors a particular allele, which then increases in frequency.

**Primates:** The order of mammals that includes prosimians, Old and New World monkeys, apes, and humans.

**Protein:** One or several chains of amino acids that have undergone complex folding.

**Proto-Indo-European:** *See* Indo-European.

**Pyramidal neuron:** A neuron with a long axon and many dendrites located in the hippocampus and cerebral cortex. Pyramidal neurons constitute approximately 80 percent of cortical neurons.

**Recessive:** An allele that causes a detectable phenotypic characteristic only when an organism has two copies. Many recessive alleles are null alleles—they do nothing. For example, blood group O of the ABO locus is the absence of a gene product. *See also* Dominant.

**Recombination:** The process of chromosomes breaking, exchanging segments, and re-forming. Also called "crossing-over." We inherit one of each pair of chromosomes from each parent. These parental chromosomes break and re-form so that we transmit chromosomes that are composed of some of each parental chromosome.

**Sarmatians:** Horse nomads who lived in southern European Russia and the eastern Balkans in late antiquity, the successors of the Scythians. They spoke an Iranian language related to the language of the Scythians. *See also* Alans; Scythians; Vandals.

**Scythians:** Horse nomads who dominated southern Russia during early classical times and spoke an Iranian language. They were closely related to the Sarmatians, their successors.

**Selective sweep:** The process by which a new variant favored by selection increases in frequency.

**SNP:** Single nucleotide polymorphism, a DNA sequence variant in which a single nucleotide differs from others in a collection of chromosomes.

**Spanish Fly:** Cantharides, a poisonous compound secreted by some beetles. If ingested, it irritates the urinary tract and causes swelling of the genitalia. It was often used as an aphrodisiac (on Louis XIV, for example) or as a poison. Today cantharides is illegal in the United States except in animal husbandry.

**Sphingolipids:** A class of lipids that are found in cell membranes and common in nerve tissue. They play a role in cell recognition, membrane structure, and signal transmission.

**Tay-Sachs disease:** A lysosomal storage disease that is unusually common (100 times the norm) in Ashkenazi Jews. Homozygotes undergo aberrant sprouting of dendrites from pyramidal neurons and die in infancy.

**Vandals:** An East Germanic tribe that invaded the Roman Empire in the fourth century. They passed through France, occupied southern Spain for a time, and later built a robber kingdom centered in North Africa. *See also* Alans; Sarmatians.

**X chromosome:** One of the two sex-determining chromosomes in mammals. Females have two copies of the X chromosome, whereas males have an X and a Y chromosome. *See also* Y chromosome.

**Yanomamo:** An Amerindian tribe of the Amazon basin that subsists by gardening and hunting. They are famous for high levels of violence and local raiding and warfare; males who have killed others have higher fitness than nonkillers.

**Y chromosome:** One of the two sex-determining chromosomes in mammals. Males have one X and one Y chromosome. The Y chromosome triggers male development. Except for mutation, it is passed on unchanged from father to son. *See also* X chromosome.

# BIBLIOGRAPHY

Achilli, Alessandro, Anna Olivieri, Maria Pala, Ene Metspalu, Simona Fornarino, Vincenza Battaglia, Matteo Accetturo et al. "Mitochondrial DNA Variation of Modern Tuscans Supports the Near Eastern Origin of Etruscans." *American Journal of Human Genetics* 80, no 4 (2007): 759–768.

Allen, Robert C. "Agriculture and the Origins of the State in Ancient Egypt." *Explorations in Economic History* 34, no. 2 (1997): 135–154.

Anderson, Roy C. "The Ecological Relationships of Meningeal Worm and Native Cervids in North America." *Journal of Wildlife Diseases* 8, no. 4 (1972): 304–310.

Ankori, Zvi. "Origins and History of Ashkenazi Jewry (8th to 18th Century)." In Richard M. Goodman and Arno G. Motulsky, eds., *Genetic Diseases among Ashkenazi Jews.* New York: Raven Press, 1979.

Anthony, David. *The Horse, the Wheel, and Language: How Bronze-Age Riders from the Eurasian Steppes Shaped the Modern World.* Princeton, N.J.: Princeton University Press, 2007.

Arkin, Marcus. *Aspects of Jewish Economic History.* Philadelphia: Jewish Publication Society of America, 1975.

Backman, Margaret E. "Patterns of Mental Abilities: Ethnic, Socioeconomic, and Sex Differences." *American Educational Research Journal* 9 (1972): 1–12.

Baker, John R. *Race.* New York: Oxford University Press, 1974.

Barkow, Jerome H., Leda Cosmides, and John Tooby. *The Adapted Mind: Evolutionary Psychology and the Generation of Culture.* New York: Oxford University Press, 1992.

Barnavi, Eli, and Miriam Eliav-Feldon. *A Historical Atlas of the Jewish People: From the Time of the Patriarchs to the Present.* New York: Knopf, 1992.

Basu, Analabha, Hua Tang, Xiaofeng Zhu, C. Charles Gu, Craig Hanis, Eric Boerwinkle, and Neil Risch. "Genome-Wide Distribution of Ancestry in Mexican Americans." *Human Genetics,* DOI 10.1007/s00439-008-0541-5.

Ben-Sasson, Hayim. *A History of the Jewish People.* Cambridge: Harvard University Press, 1976.

Bielicki, John K., Trudy M. Forte, Mark R. McCall, Lori J. Stoltzfus, Giulia Chiesa, Cesare R. Sirtori, Guido Franceschini, and Edward M. Rubin. "High Density Lipoprotein Particle Size Restriction in Apolipoprotein A-I (Milano) Transgenic Mice." *Journal of Lipid Research* 38, no. 11 (1997): 2314–2321.

Bockstette, Valerie, Areendam Chanda, and Louis Putterman. "States and Markets: The Advantage of an Early Start." *Journal of Economic Growth* 7 (2002): 347–369.

Botticini, Maristella., and Zvi Eckstein. "From Farmers to Merchants: A Human Capital Interpretation of Jewish Economic History," 2002, http://www.cepr.org/pubs/dps/DP3718.asp.

Bulliet, Richard W. *The Camel and the Wheel.* Cambridge: Harvard University Press, 1975.

Burger, J., M. Kirchner, B. Bramanti, W. Haak, and M. G. Thomas. "Absence of the Lactase-Persistence-Associated Allele in Early Neolithic Europeans." *Proceedings of the National Academy of Sciences* 104, no. 10 (2007): 3736.

Byock, Jesse L. *Viking Age Iceland.* New York: Penguin, 2001.

Carter, Richard, and Kamini N. Mendis. "Evolutionary and Historical Aspects of the Burden of Malaria." *Clinical Microbiology Reviews* 15 (2002): 564–594.

Cavalli-Sforza, L. Luca, Paolo Menozzi, and Alberto Piazza. *The History and Geography of Human Genes.* Princeton, N.J.: Princeton University Press, 1994.

Clark, Gregory. *A Farewell to Alms.* Princeton, N.J.: Princeton University Press, 2007.

Cochran, Gregory, Jason Hardy, and Henry Harpending. "Natural History of Ashkenazi Intelligence." *Journal of Biosocial Science* 38, no. 5 (2005): 659–693.

Cohen, Mark Nathan, and George J. Armelagos. *Paleopathology at the Origins of Agriculture.* Academic Press, 1984.

Cohen, Yinon, Yitchak Haberfeld, and Tali Kristal. "Ethnicity and Mixed Ethnicity: Educational Gaps among Israeli-Born Jews." *Ethnic and Racial Studies* 30, no. 5 (2007): 896–917.

Cook, Noble David. *Born to Die: Disease and New World Conquest, 1492–1650.* Cambridge: Cambridge University Press, 1998.

Coon, Carleton S. *The Origin of Races.* New York: Alfred A. Knopf, 1963.

Coop, Graham K., Kevin Bullaughey, Francesca Luca, and Molly Przeworski. "The Timing of Selection at the Human FOXP2 Gene." *Molecular Biology and Evolution* 25, no. 7 (2008): 1257.

Cooper, Morton O., and W. J. Spillman. "Farmer's Bulletin No. 877—Human Food from an Acre of Staple Farm Products." In *Farmers' Bulletin* of the U.S. Department of Agriculture. Washington, D.C.: Government Printing Office, 1919.

Cordain, L., S. B. Eaton, J. Brand Miller, N. Mann, K. Hill. "Original Communication: The Paradoxical Nature of Hunter-Gatherer Diets: Meat-Based, Yet Non-Atherogenic." *European Journal of Clinical Nutrition* 56, no. 1 (2002): S42–S52.

Crosby, Alfred W. *Ecological Imperialism: The Biological Expansion of Europe, 900–1900.* New York: Cambridge University Press, 1986.

Crotty, Raymond D. *When Histories Collide: The Development and Impact of Individualistic Capitalism.* Walnut Creek, Calif.: AltaMira Press, 2001.

Crow, James F. "Unequal by Nature: A Geneticist's Perspective on Human Differences." *Daedalus,* Winter 2002, 81–88.

Curnoe, D., A. Thorne, and J. A. Coate. "Timing and Tempo of Primate Speciation." *Journal of Evolutionary Biology* 19, no. 1 (2006): 59–65.

Currat, M., and L. Excoffier. "Modern Humans Did Not Admix with Neanderthals during Their Range Expansion into Europe." *PLoS Biology* 2, no. 12 (2004).

Darwin, Charles R. *The Voyage of the Beagle.* New York: Bantam Books, 1958.

David, Hanna, and Richard Lynn. "Intelligence Differences between European and Oriental Jews in Israel." *Journal of Biosocial Science* 39, no. 3 (2007): 465–473.

Diamond, Jared. *Guns, Germs, and Steel: The Fates of Human Societies.* New York: W. W. Norton, 2005.

———. *The Third Chimpanzee: The Evolution and Future of the Human Animal.* New York: HarperCollins, 1992.

———. "The Worst Mistake in the History of the Human Race." *Discover* 8, no. 5 (1987): 64–66.

Duarte, Cidália, João Maurício, Paul B. Pettitt, Pedro Souto, Erik Trinkaus, Hans van der Plicht, and João Zilhão. "The Early Upper Paleolithic Human Skeleton from the Abrigo do Lagar Velho (Portugal) and Modern Human Emergence in Iberia." *Proceedings of the National Academy of Sciences* 96 (1999): 7604–7609.

Dupanloup, Isabelle, Giorgio Bertorelle, Lounès Chikhi, and Guido Barbujani. "Estimating the Impact of Prehistoric Admixture on the Genome of Europeans." *Molecular Biology and Evolution* 21, no. 7 (2004): 1361–1372.

Eiberg, Hans, Jesper Troelsen, Mette Nielsen, Annemette Mikkelsen, Jonas Mengel-From, Klaus W. Kjaer, and Lars Hansen. "Blue Eye

Color in Humans May Be Caused by a Perfectly Associated Founder Mutation in a Regulatory Element Located within the HERC2 Gene Inhibiting OCA2 Expression." *Human Genetics* 123, no. 2 (2008): 177–187.

Eldridge, Roswell. "Edward Flatau, Wladyslaw Sterling, Torsion Spasm in Jewish Children, and the Early History of Human Genetics." *Advances in Neurology* 14 (1976): 105–114.

———. "Torsion Dystonias: Genetic and Clinical Studies." *Neurology* 11 (1970): 1–78.

Enattah, N. S., T. G. Jensen, M. Nielsen, R. Lewinski, M. Kuokkanen, H. Rasinpera, H. El-Shanti et al. "Independent Introduction of Two Lactase-Persistence Alleles into Human Populations Reflects Different History of Adaptation to Milk Culture." *American Journal of Human Genetics* 82, no. 1 (2008): 57–72.

Evans, P. D., S. L. Gilbert, N. Mekel-Bobrov, E. J. Vallender, J. R. Anderson, L. M. Vaez-Asizi, S. A. Tishkoff, R. R. Hudson, and B. T. Lahn. "Microcephalin, a Gene Regulating Brain Size, Continues to Evolve Adaptively in Humans." *Science* 309, no. 5741 (2005): 1717–1720.

Evans, P. D., N. Mekel-Bobrov, E. J. Vallender, R.R. Hudson, and B.T. Lahn. "Evidence That the Adaptive Allele of the Brain Size Gene Microcephalin Introgressed into *Homo sapiens* from an Archaic Homo lineage." *Proceedings of the National Academy of Sciences* 103, no. 48 (2006), DOI link.

Firasat, Sadaf, Shagufta Khaliq, Aisha Mohyuddin, Myrto Papaioannou, Chris Tyler-Smith, Peter A. Underhill, and Qasim Ayub. "Y-chromosomal Evidence for a Limited Greek Contribution to the Pathan Population of Pakistan." *European Journal of Human Genetics* 15, no. 1 (2007): 121–126.

Fix, Alan G. *Migration and Colonization in Human Microevolution.* Cambridge: Cambridge University Press, 1999.

Frayer, David W. "Evolution at the European Edge: Neanderthal and Upper Paleolithic Relationships." *Préhistoire Européenne* 2 (1993): 9–69.

———. "Perspectives on Neandertals as Ancestors." In G. A. Clark and C. M. Willermet, eds., *Conceptual Issues in Modern Human Origins Research.* New York: Aldine de Gruyter, 1998.

Fulge, M., R. Renneberg, S. Hummel, and B. Herrmann. "Lactose Persistence in Prehistoric Individuals." Eighth International Conference on Ancient DNA and Associated Biomolecules, October 16–19, 2006, Lodz, Poland.

Galvani, Alison P., and Montgomery Slatkin. "Evaluating Plague and Smallpox as Historical Selective Pressures for the CCR5-delta32

HIV-Resistance Allele." *Proceedings of the National Academy of Sciences* 100, no. 25 (2003): 15276–15279.

Garrigan, Daniel, Zahra Mobasher, Sarah B. Kingan, Jason A. Wilder, and Michael F. Hammer. "Deep Haplotype Divergence and Long-Range Linkage Disequilibrium at Xp21.1 Provides Evidence That Humans Descend from a Structured Ancestral Population." *Genetics* 170 (2005): 1849–1856.

Garrigan, Daniel, Zahra Mobasher, Tesa Severson, Jason A. Wilder, and Michael F. Hammer. "Evidence for Archaic Asian Ancestry on the Human X Chromosome." *Molecular Biology and Evolution* 22 (2005): 189–192, DOI link.

Gibbon, Edward. *A History of the Decline and Fall of the Roman Empire*, vol. 3. Philadelphia: B. F. French, 1830.

Goddard, H. "Mental Tests and the Immigrant." *Journal of Delinquency* 2 (1917): 243–277.

Gordon, Richard. *Great Medical Disasters.* New York: Stein and Day, 1983.

Gottfredson, Linda. "Logical Fallacies Used to Dismiss the Evidence on Intelligence Testing." In R. Phelps, ed., *The True Measure of Educational and Psychological Tests: Correcting Fallacies about the Science of Testing.* Washington, D.C.: American Psychological Association, in press.

———. "Why *g* Matters: The Complexity of Everyday Life." *Intelligence* 24 (1997): 79–132.

Gould, Stephen J. "The Spice of Life." *Leader to Leader* 15 (2000): 19–28.

Gregory of Tours. *The History of the Franks.* Harmondsworth, UK: Penguin, 1974.

Gross, Nachum. *Economic History of the Jews.* New York: Schocken Books, 1975.

Grousset, Rene. *The Empire of the Steppes: A History of Central Asia.* New Brunswick, N.J.: Rutgers University Press, 1970.

Gualandri, Valter, Guido Franceschini, Cesare R. Sirtori, Gemma Gianfranceschi, Giovan Battista Orsini, Antonio Cerrone, and Alessandro Menotti. "AIMilano Apoprotein Identification of the Complete Kindred and Evidence of a Dominant Genetic Transmission." *American Journal of Human Genetics* 37, no. 6 (1985).

Hamblin, Martha T., and Anna Di Rienzo. "Detection of the Signature of Natural Selection in Humans: Evidence from the Duffy Blood Group Locus." *American Journal of Human Genetics* 66, no. 5 (2000): 1669–1679.

Hamblin, Martha T., Emma E. Thompson, and Anna Di Rienzo. "Complex Signatures of Natural Selection at the Duffy Blood

Group Locus." *American Journal of Human Genetics* 70 (2002): 369–383.

Hamilton, William D. *Narrow Roads of Gene Land: The Collected Papers of W. D. Hamilton.* Vol. 1, *Evolution of Social Behaviour.* New York, Oxford University Press, 1998.

Hardy, J., A. Pittman, A. Myers, K. Gwinn-Hardy, H. C. Fung, R. de Silva, M. Hutton, and J. Duckworth. "Evidence Suggesting That *Homo neanderthalensis* Contributed the H2 MAPT Haplotype to *Homo sapiens.*" *Biochemical Society Transactions* 33 (2005): 582–585.

Hare, Brian, Michelle Brown, Christina Williamson, and Michael Tomasello. "The Domestication of Social Cognition in Dogs." *Science* 298 (2002): 1634–1636.

Harpending, Henry, and Jay Sobus. "Sociopathy as an Adaptation." *Ethology and Sociobiology* 8, no. 3 (1987): 63–72.

Harper, Kristin N., Paolo S. Ocampo, Bret M. Steiner, Robert W. George, Michael S. Silverman, Shelly Bolotin, Allan Pillay, Nigel J. Saunders, and George J. Armelagos. "On the Origin of the Treponematoses: A Phylogenetic Approach." *PLoS Neglected Tropical Diseases* 2, no. 1: e148 doi:10.1371/journal.pntd.0000148.

Hawks, John. "Adaptive Evolution of Human Hearing and the Appearance of Language." Seventy-Seventh Annual Meeting of the American Association of Physical Anthropologists, April 11, 2008, Columbus, Ohio.

Hawks, John, and Gregory Cochran. "Dynamics of Adaptive Introgression from Archaic to Modern Humans." *PaleoAnthropology* (2006): 101–115.

Hawks, John, Eric T. Wang, Gregory M. Cochran, Henry C. Harpending, and Robert K. Moyzis. "Recent Acceleration of Human Adaptive Evolution." *Proceedings of the National Academy of Sciences* 104, no. 52 (2007): 20753.

Hawks, John, and M. H. Wolpoff. "The Accretion Model of Neandertal Evolution." *Evolution* 55 (2001): 1474–1485.

Helgason, A., S. Pálsson, G. Thorleifsson, S. F. Grant, V. Emilsson, S. Gunnarsdottir, A. Adeyemo et al. "Refining the Impact of TCF7L2 Gene Variants on Type 2 Diabetes and Adaptive Evolution." *Nature Genetics* 39 (2007): 218–225.

Henshilwood, Christopher S., Francesco d'Errico, Royden Yates, Zenobia Jacobs, Chantal Tribolo, Geoff A. T. Duller, Norbert Mercier et al. "Emergence of Modern Human Behavior: Middle Stone Age Engravings from South Africa." *Science* 295, no. 5558 (2002): 1278–1280.

Hibbs, Douglas O., and Ola Olsson. "Geography, Biogeography, and Why Some Countries Are Rich and Others Are Poor." *Proceedings of the National Academy of Sciences* 101 (2004): 3715–3720.

Holliday, Trenton W. "Speciation by Distance and Temporal Overlap: A New Approach to Understanding Neanderthal Evolution." In T. Harrison and K. Harvati, eds., *Neanderthals Revisited: New Approaches and Perspectives*. New York: Sinauer, 2006.

Hoodbhoy, P. A. "Science and the Islamic World—The Quest for Rapprochement." *Physics Today* 60, no. 8 (2007): 49–55.

Hudjashov, Georgi, Toomas Kivisild, Peter A. Underhill, Phillip Endicott, Juan J. Sanchez, Alice A. Lin, Peidong Shen et al. "Revealing the Prehistoric Settlement of Australia by Y-Chromosome and mtDNA Analysis." *Proceedings of the National Academy of Sciences* 104, no. 21 (2007).

Hughes, A. "Jews and Gentiles: Their Intellectual and Temperamental Differences." *Eugenics Review* 18 (July 1928): 1–6.

Hundert, G. D. *The Jews in a Polish Private Town: The Case of Opatów in the Eighteenth Century*. Baltimore: Johns Hopkins University Press, 1992.

Hurtado, Ana Magdalena, Kim Hill, Hillard Kaplan, and Jane Lancaster. "The Epidemiology of Infectious Diseases among South American Indians: A Call for Guidelines for Ethical Research." *Current Anthropology* 42, no. 3 (2001): 425–432.

Jamison, Kay R. *Touched with Fire: Manic-Depressive Illness and the Artistic Temperament*. New York: Free Press, 1993.

Jensen, Arthur R. *Bias in Mental Testing*. New York: Free Press, 1980.

Kamen, Henry. *Empire: How Spain Became a World Power, 1492–1763*. New York: HarperCollins, 2003.

Kamin, Leon. *The Science and Politics of IQ*. Potomac, Md.: Erlbaum, 1974.

Keeley, Lawrence H. *War before Civilization: The Myth of the Peaceful Savage*. New York: Oxford University Press, 1996.

Keller, Laurent, and Kenneth G. Ross. "Selfish Genes: A Green Beard in the Red Fire Ant." *Nature* 394 (1998): 573.

Klein, Richard. *The Dawn of Human Culture*. New York: Wiley, 2002.

———. *The Human Career*. Chicago: University of Chicago Press, 1999.

Konner, Mel. *Unsettled: An Anthropology of the Jews*. New York: Viking Compass, 2003.

Krause, Johannes, Carles Lalueza-Fox, Ludovic Orlando, Wolfgang Enard, Richard E. Green, Hernán A. Burbano, Jean-Jacques Hublin et al. "The Derived FOXP2 Variant of Modern Humans Was Shared with Neandertals." *Current Biology* 17, no. 21 (2007): 1908–1912.

Krieger, Michael J. B., and Kenneth G. Ross. "Identification of a Major Gene Regulating Complex Social Behavior." *Science* 295, no. 5553 (2002): 328–332.

Larsen, Clark. *Bioarchaeology: Interpreting Behavior from the Human Skeleton.* New York: Cambridge University Press, 1999.

Levinson, Boris. "A Comparison of the Performance of Monolingual and Bilingual Native-Born, Jewish Preschool Children of Traditional Parentage on Four Intelligence Tests." *Journal of Clinical Psychology* 15 (1959): 74–76.

Lewis, Bernard. *The Jews of Islam.* Princeton, N.J.: Princeton University Press, 1984.

Lewontin, Richard. "The Apportionment of Human Diversity." *Evolutionary Biology* 6, no. 1 (1972): 381–398.

Lincoln, Bruce. *Priests, Warriors, and Cattle: A Study in the Ecology of Religions.* Berkeley: University of California Press, 1981.

Lynn, Richard. "The Intelligence of American Jews." *Personality and Individual Differences* 26 (2004): 201–206.

MacArthur, Daniel G., Jane T. Seto, Joanna M. Raftery, Kate G. Quinlan, Gavin A. Huttley, Jeff W. Hook, Frances A. Lemckert et al. "Loss of ACTN3 Gene Function Alters Mouse Muscle Metabolism and Shows Evidence of Positive Selection in Humans." *Nature Genetics* 39, no. 10 (2007).

Maier, A. G., M. T. Duraisingh, J. C. Reeder, S. S. Patel, J. W. Kazura, P. A. Zimmerman, and A. F. Cowman. "*Plasmodium falciparum* Erythrocyte Invasion through Glycophorin C and Selection for Gerbich Negativity in Human Populations." *Nature Medicine* 9, no. 1 (2003) 87–92.

Mallory, James P. *In Search of the Indo-Europeans: Language, Archaeology and Myth.* London: Thames and Hudson, 1989.

Mallory, James P., and Douglas Q. Adams. *The Oxford Introduction to Proto-Indo-European and the Proto-Indo-European World.* New York: Oxford University Press, 2006.

Martinez-Marignac, Veronica L., Adan Valladares, Emily Cameron, Andrea Chan, Arjuna Perera, Rachel Globus-Goldberg, Niels Wacher et al. "Admixture in Mexico City: Implications for Admixture Mapping of Type 2 Diabetes Genetic Risk Factors." *Human Genetics* 120, no. 6 (2007): 807–819.

Mayr, E. *Animal Species and Evolution.* Cambridge: Belknap Press of Harvard University Press, 1963.

McNeill, John Robert, and William H. McNeill. *The Human Web: A Bird's-Eye View of World History.* New York: W. W. Norton, 2003.

McNeill, William H. *Plagues and Peoples.* Garden City, N.Y.: Anchor Press/Doubleday, 1976.

Modell, Bernadette, and Matthew Darlison. "Global Epidemiology of Haemoglobin Disorders and Derived Service Indicators." *Bulletin of the World Health Organization* 86, no. 6 (2008): 480–487.

Moore, Laoise T., Brian McEvoy, Eleanor Cape, Katharine Simms, and Daniel G. Bradley. "A Y-Chromosome Signature of Hegemony in Gaelic Ireland." *American Journal of Human Genetics.* 78, no. 2 (2006): 334–338.

Mosher, Dana S., Pascale Quignon, Carlos D. Bustamante, Nathan B. Sutter, Cathryn S. Mellersh, Heidi G. Parker, Elaine A. Ostrander et al. "A Mutation in the Myostatin Gene Increases Muscle Mass and Enhances Racing Performance in Heterozygote Dogs." *PLoS Genetics* 3, no. 5 (2007): e79.

Neel, James. "Diabetes Mellitus: A 'Thrifty' Genotype Rendered Detrimental by 'Progress'?" *American Journal of Human Genetics* 14 (1962): 353–362.

Nesse, R. M., and George C. Williams. "Darwinian Medicine." *Life Science Research* 3 (1999): 1–17.

New, Maria I., and R. C. Wilson. "Steroid Disorders in Children: Congenital Adrenal Hyperplasia and Apparent Mineralocorticoid Excess." *Proceedings of the National Academy of Sciences of the USA* 96 (1999): 12790–12797.

O'Connell, James, and James Allen. "Dating the Colonization of Sahul (Pleistocene Australia—New Guinea): A Review of Recent Research." *Journal of Archaeological Science* 31, no. 6 (2004): 835–853.

Ohashi, Jun, Izumi Naka, Jintana Patarapotikul, Hathairad Hananantachai, Gary Brittenham, Sornchai Looareesuwan, Andrew G. Clark, and Katsushi Tokunaga. "Extended Linkage Disequilibrium surrounding the Hemoglobin E Variant due to Malarial Selection." *American Journal of Human Genetics* 74, no. 6 (2004): 1198–1208.

Olshen, Adam B., Bert Gold, Kirk E. Lohmueller, Jeffrey P. Struewing, Jaya Satagopan, Stefan A. Stefanaov, Eliazar Eskin et al. "Analysis of Genetic Variation in Ashkenazi Jews by High Density SNP Genotyping." *BMC Genetics* 9, no. 1 (2008): 14.

Patai, Raphael, and Jennifer Patai. *The Myth of the Jewish Race.* New York: Scribner, 1975.

Pellecchia, Marco, Riccardo Negrini, Licia Colli, Massimiliano Patrini, Elisabetta Milanesi, Alessandro Achilli, Giorgio Bertorelle et al. "The Mystery of Etruscan Origins: Novel Clues from *Bos Taurus* Mitochondrial DNA." *Proceedings of the Royal Society B: Biological Science* 274, no. 1614 (2007): 1175–1179.

Perry, George H., Nathaniel J. Dominy, Katrina G. Claw, Arthur S. Lee, Heike Fiegler, Richard Redon, John Werner et al. "Diet and the Evolution of Human Amylase Gene Copy Number Variation." *Nature Genetics* 39 (2007): 1256–1260.

Plagnol, Vincent, and Jeffrey D. Wall. "Possible Ancestral Structure in Human Populations." *PLoS Genetics* 2 (2006): e105, DOI link.

Protas, Meredith E., Candace Hersey, Dawn Kochanek, Yi Zhou, Horst Wilkens, William R. Jeffery, Leonard I. Zon, Richard Borowsky, and Clifford J. Tabin. "Genetic Analysis of Cavefish Reveals Molecular Convergence in the Evolution of Albinism." *Nature Genetics* 38, no. 1 (2006): 107–111.

Reader, John. *Africa: A Biography of the Continent.* New York: Knopf, 1998.

Richards, Martin. "The Neolithic Invasion of Europe." *Annual Review of Anthropology* 32, no. 1 (2003): 135–162.

Rieger, Mark. *Introduction to Fruit Crops.* New York: Food Products Press, 2006.

Rock, W. P., A. M. Sabieha, et al. "A Cephalometric Comparison of Skulls from the Fourteenth, Sixteenth and Twentieth Centuries." *British Dental Journal* 200 (2006): 33–37.

Romanoff, Julius S. "Birth Order, Family Size, and Sibling Spacing as Influences on Intelligence and Academic Abilities of Jewish Adolescents." Department of Psychology, Temple University, 1976.

Russell, Cyril, and Harry S. Lewis. *The Jew in London.* London: Harper Collins, 1900.

Sallares, Robert, Abigail Bouwman, and Cecilia Anderung. "The Spread of Malaria to Southern Europe in Antiquity: New Approaches to Old Problems." *Medical History* 48, no. 3 (2004): 311–328.

Saunders, Matthew A., Montgomery Slatkin, Chad Garner, Michael F. Hammer, and Michael W. Nachman. "The Extent of Linkage Disequilibrium Caused by Selection on G6PD in Humans." *Genetics* 171, no. 3 (2005): 1219–1229.

Schultz, Jack C., and Ted Floyd. "Desert Survivor." *Natural History* 108, no. 2 (1999): 24–29.

Schwartz, Andreas, Elizabeth Rapaport, Koret Hirschberg, and Anthony H. Futerman. "A Regulatory Role for Sphingolipids in Neuronal Growth: Inhibition of Sphingolipid Synthesis and Degradation Have Opposite Effects on Axonal Branching." *Journal of Biological Chemistry* 270, no. 18 (1995): 10990–10998.

Seligman, Daniel. *A Question of Intelligence: The IQ Debate in America.* New York: Birch Lane Press, 1992.

Serre, D., A. Langaney, M. Chech, M. Teschler-Nicola, M. Paunovic, P. Mennecier, M. Hofreiter, G. Possnert, and S. Paabo. "No Evidence of Neandertal mtDNA Contribution to Early Modern Humans." *PLoS Biology* 2, no. 3 (2004): 313–317.

Sirtori, Cesare R., Laura Calabresi, Guido Franceschini, Damiano Baldassarre, Mauro Amato, Jan Johansson, Massimo Salvetti et al. "Cardiovascular Status of Carriers of the Apolipoprotein A-IMilano Mutant." *Circulation* 103, no. 15 (2001): 1949–1954.

Slatkin, Montgomery, and Christina A. Muirhead. "A Method for Estimating the Intensity of Overdominant Selection from the Distribution of Allele Frequencies." *Genetics* 156, no. 4 (2000): 2119–2126.

Smith, John Maynard. *Evolution and the Theory of Games.* Cambridge: Cambridge University Press, 1982.

Stevens, Gwynneth, Michèle Ramsay, and Trefor Jenkins. "Oculocutaneous Albinism (OCA2) in Sub-Saharan Africa: Distribution of the Common 2.7-kb P Gene Deletion Mutation." *Human Genetics* 99, no. 4 (1997): 523–527.

Stringer, Chris, and Peter Andrews. *The Complete World of Human Evolution.* London: Thames and Hudson, 2005.

———. "Genetic and Fossil Evidence for the Origin of Modern Humans." *Science* 239, no. 4845 (1988): 1263–1268.

Stutz, H. C., and L. K. Thomas. "Hybridization and Introgression in Cowania and Purshia." *Evolution* 18, no. 2 (1964): 183–195.

Sweeney, J., and T. Samansky. "Elements of Successful Facility Design: Marine Mammals." In E. F. Gibbons, B. Durrant, and J. Demarest, eds., *Conservation of Endangered Species in Captivity: An Interdisciplinary Approach.* New York: State University of New York Press, 1995.

Tedlow, Richard S., Courtney Purrington, and Kim Eric Bettcher. "The American CEO in the Twentieth Century: Demography and Career Path." Harvard NOM Working Paper No. 03-21, Harvard Business School Working Paper No. 03-097. February 2003. Available at SSRN: http://ssrn.com/abstract=383280 or DOI: 10.2139/ssrn.10.2139/ssrn.38328.

Templeton, A. R. "Haplotype Trees and Modern Human Origins." *Yearbook of Physical Anthropology* 48 (2005): 33–59, DOI link.

Thomas, Mark G., Michael P. H. Stumpf, and Heinrich Härke. "Evidence for an Apartheid-Like Social Structure in Early Anglo-Saxon England." *Proceedings of the Royal Society B: Biological Sciences* 273, no. 1601 (2006): 2651–2657.

Thompson, E. E., H. Kuttab-Boulos, D. Witonsky, L. Yang, B. A. Roe, and A. Di Rienzo. "CYP3A Variation and the Evolution of Salt-Sensitivity Variants." *American Journal of Human Genetics* 75, no. 6 (2004): 1059–1069.

Tian, Chao, Robert M. Plenge, Michael Ransom, Annette Lee, Pablo Villoslada, Carlo Selmi, Lars Klareskog et al. "Analysis and Application of European Genetic Substructure Using 300K SNP Information." *PLoS Genetics* 4, no. 1 (2008): e4.

Tishkoff, Sarah A., Floyd A. Reed, Alessia Ranciaro, Benjamin F. Voight, Courtney C. Babbitt, Jesse S. Silverman, Kweli Powell et al. "Convergent Adaptation of Human Lactase Persistence in Africa and Europe." *Nature Genetics* 39, no. 1 (2007): 31–40.

Tompkins, Daniel M., Anthony W. Sainsbury, Peter Nettleton, D. Buxton, and J. Gurnell. "Parapoxvirus Causes a Deleterious Disease in Red Squirrels Associated with UK Population Declines." *Proceedings of the Royal Society, B: Biological Sciences* 269, no. 1490 (2002).

Tooby, John, and Leda Cosmides. "On the Universality of Human Nature and the Uniqueness of the Individual: The Role of Genetics and Adaptation." *Journal of Personality* 58, no. 1 (1990): 17–67.

Trinkaus, Eric. "Early Modern Humans." *Annual Review of Anthropology* 34 (2005): 207–230.

Trut, Lyudmila N. "Early Canid Domestication: The Farm-Fox Experiment." *American Scientist* 161 (1999): 161.

Vervaecke, Hilde, and Linda Van Elsacker. "Hybrids between Common Chimpanzees (Pan troglodytes) and Pygmy Chimpanzees (Pan paniscus) in Captivity." *Mammalia* (Paris) 56, no. 4 (1992): 667–669.

Voight, Benjamin F., Sridhar Kudaravalli, Xiaoquan Wen, Jonathan K. Pritchard. "A Map of Recent Positive Selection in the Human Genome." *PLoS Biology* 4, no. 3 (2006): e72.

Vullaimy, T. J., M. D'Urso, G. Battistuzzi, M. Estrada, N. S. Foulkes, G. Martini, V. Calabro et al. "Diverse Point Mutations in the Human Glucose-6-Phosphate Dehydrogenase Gene Cause Enzyme Deficiency and Mild or Severe Hemolytic Anemia." *Proceedings of the National Academy of Sciences* 85, no. 14 (1988): 5171–5175.

Walkley, Steven U. "Neurobiology and Cellular Pathogenesis of Glycolipid Storage Diseases." *Philosophical Transactions of the Royal Society London B* 358 (2003): 893–904.

Walkley, Steven U., Mark Zervas, and Samson Wiseman. "Gangliosides as Modulators of Dendritogenesis in Normal and Storage Disease-Affected Pyramidal Neurons." *Cerebral Cortex* 10 (2000): 1028–1037.

Wang, Eric T., Greg Kodama, Pierre Baldi, and Robert K. Moyzis. "Global Landscape of Recent Inferred Darwinian Selection for *Homo sapiens*." *Proceedings of the National Academy of Sciences* 103, no. 1 (2006): 135–140.

Weaver, T. D., and C. C. Roseman. "Ancient DNA, Late Neandertal Survival, and Modern-Human–Neandertal Genetic Admixture." *Current Anthropology* 46, no. 4 (2005): 677–683.

Weinryb, Bernard D. *The Jews of Poland: A Social and Economic History of the Jewish Community in Poland from 1100 to 1800.* Philadelphia: Jewish Publication Society of America, 1973.

Williams, Margaret H. *The Jews among the Greeks and Romans: A Diasporan Sourcebook.* Baltimore: Johns Hopkins University Press, 1998.

Williamson, Scott H., Melissa J. Hubisz, Andrew G. Clark, Bret A. Payseur, Carlos D. Bustamante, and Rasmus Nielsen. "Localizing Recent Adaptive Evolution in the Human Genome." *PLoS Genetics* 10 (2007).

Wolpoff, Milford, John Hawks, David Frayer, and Keith Hunley. "Modern Human Ancestry at the Peripheries: A Test of the Replacement Theory." *Science* 291, no. 5502 (2001): 293–297.

Xue, Yali, Allan Daly, Bryndis Yngvadottir, Mengning Liu, Graham Coop, Yuseob Kim, Pardis Sabeti et al. "Spread of an Inactive Form of Caspase-12 in Humans Is Due to Recent Positive Selection." *American Journal of Human Genetics* 78, no. 4 (2006): 659–670.

Yi, Zanhua, Nanibaa Garrison, Orit Cohen-Barak, Tatiana M. Karafet, Richard A. King, Robert P. Erickson, Michael F. Hammer, and Murray H. Brilliant. "A 122.5-Kilobase Deletion of the P Gene Underlies the High Prevalence of Oculocutaneous Albinism Type 2 in the Navajo Population." *American Journal of Human Genetics* 72, no. 1 (2003): 62–72.

Yoshiura, Koh-ichiro, Akira Kinoshita, Takafumi Ishida, Aya Ninokata, Toshihisa Ishikawa, Tadashi Kaname, Makoto Bannai et al. "A SNP in the ABCC11 Gene Is the Determinant of Human Earwax Type." *Nature Genetics* 38 (2006): 324–330.

Zayed, Amos, and Charles W. Whitfield. "A Genome-Wide Signature of Positive Selection in Ancient and Recent Invasive Expansions of the Honey Bee Apis mellifera." *Proceedings of the National Academy of Sciences* 105, no. 9 (2008).

Zegura, Stephen L., Tatiana M. Karafet, Lev A. Zhivotovsky, and Michael F. Hammer. "High-Resolution SNPs and Microsatellite Haplotypes Point to a Single, Recent Entry of Native American Y Chromosomes into the Americas." *Molecular Biology and Evolution* 21, no. 1 (2004): 164–175.

Zerjal, Tatiana, Yali Xue, Giorgio Bertorelle, R. Spencer Wells, Weidong Bao, Suling Zhu, Raheel Qamar et al. "The Genetic Legacy of the Mongols." *American Journal of Human Genetics* 72, no. 3 (2003): 717–721.

Zietkiewicz, E., V. Yotova, D. Gehl, T. Wambach, I. Arrieta, M. Batzer, D. E. Cole et al. "Haplotypes in the Dystrophin DNA Segment

Point to a Mosaic Origin of Modern Human Diversity." *American Journal of Human Genetics* 73 (2003): 994–1015.

Zohary, Daniel, and Maria Hopf. *Domestication of Plants in the Old World: The Origin and Spread of Cultivated Plants in West Asia, Europe, and the Nile Valley.* New York: Oxford University Press, 2001.

# CREDITS

xiii [Timeline] Henry Harpending, University of Utah
6 [Wolf] U.S. Fish and Wildlife Service, Department of the Interior
6 [Two dogs] Deanne Fitzmaurice
8 [Teosinte and corn] National Science Foundation
35 [Lascaux] Larry Dale Gordon, Getty Images
38 [Venus of Dolni Vestonice] Petr Novák, Wikipedia, Creative Commons Attribution ShareAlike 2.5
38 [Venus of Willendorf] Matthias Kabel, Wikipedia, Creative Commons CC-BY 2.5
38 [Blombos ochre] Christopher Henshilwood, National Science Foundation
39 [Lion Man of Hohlenstein] John Hawks, University of Wisconsin
47 [AIDS retrovirus] Los Alamos National Laboratories
49 [Zebu cow] U.S. Department of Agriculture
49 [Texas longhorn] Larry D. Moore, Wikipedia, Creative Commons Attribution ShareAlike 2.5
110 [Man with the Hoe] The J. Paul Getty Museum, Los Angeles. Jean-François Millet, Man with a Hoe, 1862, black chalk and white chalk heightening on buff paper, 28.1 × 34.9 cm ($11\frac{1}{16} \times 13\frac{3}{4}$ in.).
133 [Limone sul Garda] Corbis
150 [Tuareg] Florence Devouard, Wikipedia, Creative Commons Attribution ShareAlike 1.0
151 [Afghan girl] Steve McCurry, Magnum Photos
192 [Two bell curves] Henry Harpending, University of Utah
205 [European genetic substructure analysis] Chao Tian, Robert M. Plenge, Michael Ransom, Annette Lee, Pablo Villoslada, Carlo Selmi, Lars Klareskog et al., "Analysis and Application of European Genetic Substructure Using 300K SNP Information," PLOS Genetics 4, no. 1 (2008): e4. Creative Commons Attribution 2.5 Generic.
218 [Whippets] Dana S. Mosher, Pascale Quignon, Carlos D. Bustamante, Nathan B. Sutter, Cathryn S. Mellersh, Heidi G. Parker,

Elaine A. Ostrander et al., "A Mutation in the Myostatin Gene Increases Muscle Mass and Enhances Racing Performance in Heterozygote Dogs," PLOS Genetics 3, no. 5 (2007): e79. Creative Commons Attribution 2.5 Generic.

221 [drawing of neurons showing axon growth] Andreas Schwarz, Elizabeth Rapaport, Koret Hirschberg, and Anthony H. Futerman, "A Regulatory Role for Sphingolipids in Neuronal Growth: Inhibition of Sphingolipid Synthesis and Degradation Have Opposite Effects on Axonal Branching," *Journal of Biological Chemistry* 270, no. 18 (1995): 10990–10998.

# INDEX


Achondroplasia, 133
Adaptive traits, 243
Adret, Rabbi Solomon ben Abraham, 201–202
Afghan girl, 151 (photo)
Africa
  fossilization and, 62
  lactose tolerance, 77–78, 139, 185
  out-of-Africa expansion, 3, 14, 25, 26, 28, 31, 36, 63, 64, 130, 139, 156, 156–157, 171–173, 225–226
  slavery, 139, 141, 153, 171
  *See also specific groups of people*; Sub-Saharan Africa
African/Africanized bees, 51
Agobard of Lyons, 195
Agriculture transition
  beginnings/spread, 68–69, 79
  changes overview, 69–71
  cultural change, 84
  demic expansion, 181–182
  difficulties with (overview), 59, 66
  domestication of animals, 68
  domestication of plants, 6, 8 (fig.), 13, 68
  driving genes/alleles, 96–97
  evolutionary response overview, 65–67
  farmer/hunter-gatherer personality traits, 113–118, 119
  food production/nutrition, 69–70
  game animals decline, 68
  genetic response, 71–76
  gratification deferral, 114, 117
  Holocene developments, 31
  Ice Age end, 67
  infectious disease, 69, 70, 85–91
  innovations with, 69
  intelligence, 67–68
  Malthusian trap, 100–102, 103, 104–105, 117, 118
  *Man with the Hoe* (Millet), 110 (fig.)
  mutations and, 59, 66, 74–76
  nonproductive elites, 70, 105, 111–112
  overview, 65–67
  population growth, 66, 69, 74, 226
  property, 70, 114–115, 116–117
  selective pressures, 70–71
  selective sweeps, 157
  taming/domesticating people, 111–112, 113
  work difficulty, 115, 116

Agriculture transition/diet changes
alcohol, 79
carbohydrates/disease, 76, 78–79, 81–83
cultural change, 84
diabetes, 79, 80–81, 82
genetic adaptation differences, 79–81, 82–83
genetic adaptations, 77–79, 82–84
health problems, 76–77
height changes, 76
hunter-gatherer diet vs., 76
milk, 77
pathology in bones, 77
protein, 76
vitamin D, 78
vitamin-deficiency diseases, 76
*See also* Lactase; Lactose
AIDS/HIV, 46, 47 (fig.)
Akkadian Empire, 141–142, 243
Al-Rashid, Haroun, 195
Alans, 152, 243
Albinism, 92–93
Alcohol
adaptations to, 79, 82–83
agriculture transition, 79
alleles reducing risk, 79, 82
FAS, 83
health advantages, 82
pregnant women, 83
Alcoholism
alleles reducing risk, 79, 82
diabetes, 82
Alexander the Great, 144–145
Alleles
adaptive variants, 130
advantageous alleles, 41–42, 43–44, 53, 59, 62, 64, 130
blood types, 243
definition/description, 243
dominant allele, 245–246
driving genes/alleles, 95–98
neutral alleles effects, 40–41, 42–43
probabilities in becoming common, 41–44
recessive allele, 250
Almonds, wild vs. domesticated, 17
Alpha-1-antitrypsin deficiency, 165
Alzheimer's disease, 209
A. M. Turing Awards (American Mathematical Society), 190
Amerindians
adaptations to high altitude, 56, 163
agriculture/adaptations, 68–69, 80
Altiplano of South America, 163
disease resistance weakness, 90–91, 159–170
domestic animals, 159, 168
FAS, 83
genetic history, 46, 129–130
HLA system, 160–161
killing of animal species, 159
Mexicans and, 46, 130
OCA2 gene, 93
origins, 18, 159
parasitic worm infections/immune system, 167
selection for weaker immune systems, 161
social/technical developments, 119
Western diets, 80
*See also* Columbian explosion
Amino acids, 243
Amygdalin, 17

Anatomically modern humans (AMH), 243. *See also* Modern humans
Anatomy
  Australian Aborigines, 94
  brow ridges, 94, 95
  modern humans, 26, 27, 59–60
  Neanderthals, 26, 54, 55, 59–60
  *See also* Brain development; Skeletal record; Teeth
Andaman Islands, 90–91
Arawak people, 163
Archaic humans, 243–244. *See also* Neanderthals
Aristotle, 170
Arithmetic, invention of, 180
Art
  Upper Paleolithic innovations, 30, 33, 34–35, 35 (photo)
  *See also specific types*
Arthur mythos, 148
Ashkenazi Jews
  conversions/admixture, 203, 204–205
  definition, 187
  enslavement of, 204–205
  famine/malnutrition and, 197
  finance, 196–197, 198–199
  as genetically distinct group, 203–208, 205 (fig.)
  history to 1800, 192–200
  influence of, 188, 191–192
  intermarriage prohibitions, 107–108, 193–194, 204, 205, 206, 219–220, 224
  Jewish Diaspora, 193, 194, 198–199
  Muslim world and, 195–196
  occupational patterns, 199–200, 222–223
  origins, 194–195
  overview, 107–108, 187–189
  persecution, 195, 196, 197–198, 201, 212
  Polish-Lithuanian Commonwealth, 198–200, 202
  population growth, 198, 199–200, 222
  population size, 187
  Rabbinical Judaism, 193, 194
  religious bans, 201–202
  standard of living/effects, 197, 199–200, 222
  Talmud, 193, 202
  Torah, 193, 201
  trade, 195–196
  visuospatial abilities, 212
Ashkenazi Jews/genetic diseases
  bottleneck hypothesis, 214–216
  concentration in few metabolic pathways, 213–214, 219–220
  congenital adrenal hyperplasia (CAH), 222
  DNA repair disorders, 214
  familial dysautonomia, 188, 213
  Gaucher's disease, 188, 213, 214, 221 (fig.)
  hereditary breast cancer, 188, 213, 214
  intelligence and, 220–222, 221 (fig.)
  natural selection, 216–222
  natural selection mechanism, 222–224
  Niemann-Pick disease, 213, 214, 220
  overview, 188–189
  sphingolipid storage disorders, 214, 215, 220
  summary lists, 188, 213, 214
  Tay-Sachs disease, 188, 203, 213, 214, 220, 222
  torsion dystonia, 221–222

Ashkenazi Jews/intelligence
in early history, 188, 193, 201–202, 224
emergence of intellectual differences, 192, 202–203
Enlightenment values, 202
genetic diseases and, 220–222, 221 (fig.)
as genetically distinct group, 203–208, 205 (fig.)
Goddard's study, 211–212
IQ scores, 188, 191, 210–213, 222, 223–224
legal equality, 202–203
literacy, 193
natural selection, 216–222
natural selection mechanism, 222–224
overview, 188, 189
prominence of, 188, 189–192
religious bans, 201–202
science, 188, 189–191
vs. non-Ashkenazi Jews, 212–213, 224
vs. non-Jews, 191, 210–212
Asian honey bees, 50
Assyrian Empire, 244
Assyrian forced relocations, 145–146
Atherosclerosis, 134
Atlatl darts, 32, 244
Atlatls, 3, 26, 244
Attention-Deficit/Hyperactivity Disorder (ADHD), 112
Aurignacian culture, 26, 29, 244
Austin, Thomas, 131
Australia
gene flow barriers, 138–139
rabbits, 131–132
Australian Aborigines
anatomy/brow ridges, 94
diabetes, 80
gene flow barriers, 138
as hunter-gatherers, 79
infectious diseases, 90–91, 169
Western diets, 80
Autoimmune disorders, 161
Autosome, 244
Axons, 244
Aztecs, 162, 164

Balanced behavioral polymorphisms, 72–73
Balanced polymorphisms
definition/description, 72
hawk-dove game, 72–73
Bantu, 104, 155, 156
Bar-Kochba revolt (AD 132–135), 194, 204
Barbarians and Roman Empire, 151–152
Barros, João de, 171
Basset hounds, 12
Becker's muscular dystrophy, 99
Bees. *See* Honey bees
Behavioral modernity
definition/description, 226, 244
driving genes/alleles, 96–97
Upper Paleolithic, 31
Belisarius, 152
Bell curves
with different means, 192 (fig.)
normal distribution, 124
Belloc, Hilaire, 173
Belyaev, Dmitri, 6, 112
Ben-Sasson, H., 197
Bene Israel of India, 187
Berbers, 150 (photo), 151, 153, 244
"Big bang," 32. *See also* Upper Paleolithic innovations
Birth control, 102
Blade definition/description, 244

Blombos Cave, South Africa
ochre, 35, 38 (photo)
Blue eyes, 18, 92, 148–151, 150 (photo), 151 (photo), 152, 153
Bogues, Muggsy, 209
Bonobos evolution, 37
Border collies, 10–11, 12
Bottleneck
Amish communities, 215
Ashkenazi Jews/genetic diseases, 214–216
definition/description, 214–216, 244–245
Pingelap example, 215
Polynesians, 80–81
Bow and arrow, 32
Brain development
changes overview, 98–100
domesticating people, 112
fitness, 54, 55, 57, 58
human population differences, 98–100
microcephalin (MCPH1) gene, 62–63
muscle development trade-offs, 99
Breeding experiments
overview, 52–53
trait plateau, 53
trait selection, 52
Bronze Age, 65, 84, 95, 121, 139, 142, 177, 182, 244
Brow ridges, 94, 95
Bubonic plague, 86, 161
Burton, Sir Richard Francis, 172
Bushmen
Bantu expansion, 155, 156
description, 3–4
government, 105
marital distance, 136–137
personality traits, 114
tools/weapons, 3–4
way of life, 3–4, 105, 136–137

Canavan disease, 213
Cannibalism, 81, 103
Carbon-14 dating, 245
Caribbean Islands, 163
Carolingian kings, 195
Carrying capacity
climatic changes, 103
definition/description, 245
expansion, 181
hunter-gatherers, 81
peace/violence, 81, 103, 104, 116
Carthage, Tunisia, 143
Cave paintings, 30, 34, 35 (photo)
Centromeres, 97, 245
Charlemagne, 195
Charles II, 168–169
Châtelperronian tradition, 29, 156, 245
Cheaters in society, 56, 120
Chicken pox
critical community size, 86–87
shingles, 86
Chihuahuas, 7 (photo), 13, 15, 16, 52
Chimpanzees, 22, 37
Cholesterol mutation example, 133–135
Chromosomes, 20, 245
Clark, Gregory, 104
Codon, 245
Colonization
attempts in Africa, 139, 171–173
colonizers, 142–143
gene flow, 142–144
Columbian explosion
Amerindian vs. European numbers, 162–163

Columbian explosion *(continued)*
biological differences, 158–170
disease killing adults/effects, 167–168
disease resistance differences, 159–170
disease resistance of Amerindians, 90–91, 159–170
diseases introduced (summary), 161–162
European cultural advantages, 164
European iron/steel, 164
oppression of Amerindians, 169
reduction in Amerindian populations, 162, 165, 167
smallpox, 161, 162, 165–166, 167, 169
Congenital adrenal hyperplasia (CAH), 222
Connexin-26 deafness, 165
Convergent evolution, 60
"Cootie" theory, 28
Copernicus, 125
Cortés, Hernán, 162, 227
Cosmides, Leda, 9–10
Cows, 48, 49 (photo). *See also* Lactase; Lactose; Taurine cattle; Texas longhorn; Zebu cattle
Crafoord Prize, 191
Creativity and fitness, 126–127
Creosote bush/specialized insects, 8–9
Cro-Magnons, 60
Crusade, First (1096), 197–198
"Cultural explosion," 3, 32. *See also* Upper Paleolithic innovations
Culture
genetic change vs., 2, 121–122
natural selection and, 3–4
*See also* Upper Paleolithic innovations
Cyrus the Great, 146
Cystic fibrosis, 165

Dagobert, King, 195
Darius the Great, 182, 183
Darwin, Charles, 125, 169
Davis, Sammy, Jr., 203
*De revolutionibus* (Copernicus), 125
De Soto, Hernando, 169
Deer/moose and parasites, 28
Dendrites, 245
Dengue fever, 161–162
Diabetes
agriculture adaptation, 80
alcoholism, 82
autoimmune disorders, 161
gene variants regulating, 79
high-carbohydrate diet, 79
physical activity effects, 80
Polynesians, 80–81
population size/protective mutations, 80–81
Diamond, Jared, 37, 66, 121, 140–141
Dio, Cassius, 146
Diphtheria, 161
Diploid organisms, 245
Diseases
carbohydrates and, 79
cities, 108–109
European medicine (1600s), 168–169
Malthusian trap, 100, 103, 104
recessive genetic diseases/resistance, 165
vitamin D deficiency, 78
vitamin-deficiency diseases, 76, 78

*See also* Ashkenazi Jews/genetic diseases; *specific diseases*
Diseases, infectious
agriculture transition, 69, 70, 85–91
colonization attempts in Africa, 139, 171–173
"cootie" theory, 28
critical community size, 86–87
differentiating populations, 87–91
farm animals, 86, 87
farmers vs. hunter-gatherers, 87
as gene flow barrier, 139
isolated populations, 90–91
Malthusian trap, 100, 103, 104
Old World vs. New World, 90–91, 161–162
population contact, 86, 109
population density, 86, 108–109
recessive genetic disease and, 165
rodents, 86
trade effects, 87
*See also* Columbian explosion; *specific diseases*
DNA
definition/description, 17, 245
functional percentage, 132
retroviruses, 46–47
viruses, 46
DNA repair disorders, 214
Dogs/domestication
behavioral changes, 6, 10, 11, 13, 55
breed differences, 11, 12
diversity, 5, 7 (photo), 10–11, 12, 13, 15, 16, 52
neoteny, 11, 113
"reading" people, 6, 55
time of changes, 5, 6
wolves and, 5, 6, 7 (photo), 10–11, 13, 52, 55
*See also specific breeds*
Dolni Vestonice, 34
Domesticating people, 111–113
Domestication of animals
cows, 48
horses, 176, 179, 180
physical changes, 112
time period, 5–6, 8, 68
*See also* Dogs/domestication
Domestication of plants
almonds, 17
Bantu, 156
corn/maize, 6, 8 (fig.), 13
time period, 6, 8, 68
Dominant allele. *See under* Alleles
Dowries, 116
Driving genes/alleles
centromeres and, 97
fixation, 96, 97
overview, 95–98
problems with, 97–98
rate/sweeps of, 96–97
Duchenne muscular dystrophy, 88, 99
Duffy mutation, 13
Duke of Wellington, 170
Dumézil, George, 176
Dwarfism, 133
Dystrophin, 99
Dystrophin complex, 99

Earwax, dry, 18
Eemian interglacial
definition/description, 246
limits during, 31
time of, 31, 246
Egg production, 96, 97, 98
Egyptian trade, 141
Einstein, Albert, 190
Eldridge, Roswell, 221–222

Elephant size changes, 9
Elites
  agriculture transition, 70, 105, 111–112
  curbing growth of, 105
  definition, 105
  disease and, 108, 110
  examples, 106–108
  government, 109–110, 111–112
  immunity to famine, 108
  reproductive advantage, 104–105, 106–107
  silphium use, 109
  taming/domesticating people, 111–112, 113
Elopi tribespeople, 140–141
"Emotional intelligence," 208–209
Endogamous groups
  Ashkenazi Jews example, 107–108, 193–194, 204, 205, 206, 219–220, 224
  definition/description, 204, 205
*Essay on the Principle of Population, An* (Malthus), 100
Etruscans
  as colonizers, 142–143, 144
  definition/description, 144, 246
  language, 142, 144
  origins, 144
European medicine (1600s), 168–169
Evans, P. D., 62
Evolution of humans, recent
  "as shallow," 12–13
  changes as on, off, selective, 12–13
  correlations between genetic differences, 15–16
  haplotype evidence, 21–22, 149, 151
  rate, 1, 5, 18–20, 22–23, 226–227
  significance of appearance differences, 14, 17–18
  stasis beliefs, 2, 227
  *See also specific events*
Expansions
  admixture, 155
  biological advantages/inequalities, 156–158
  cultural advantages, 156
  demic expansion, 181–182
  examples, 155–156
  out of Africa, 3, 14, 25, 26, 28, 31, 36, 63, 64, 130, 139, 156, 156–157, 171–173, 225–226
  overview, 155–158
  Proto-Indo-Europeans, 174, 177–186
  replacement, 155, 156
  *See also* Columbian explosion; Modern humans displacement of Neanderthals
Exponential growth
  definition/description, 246
  examples, 66, 75, 97
Eye color
  blue eyes, 18, 92, 148–151, 150 (photo), 151 (photo), 152, 153
  OCA2 allele, 18, 149–151, 152, 153
  rate of change, 18
  variety, 94

Familial dysautonomia, 188, 213
Familial Mediterranean fever, 165
Famine
  cannibalism, 81
  hunter-gatherers and, 81–82

Malthusian trap, 100, 102, 103, 104
northern Europe (1315–1317), 81
*Farewell to Alms, A* (Clark), 104
FAS. *See* Fetal alcohol syndrome
Fayu tribespeople, 140–141
Fertile Crescent, Southwest Asia, 68
Fetal alcohol syndrome (FAS), 83
Feynman, Richard, 190–191
Fields Medal, 191
First Crusade (1096), 197–198
Fishes in caves
albinism, 93
eyesight loss, 11–12, 92–93
Fitness
brain development and, 54, 55, 57, 58
creativity and, 126–127
definition/description, 246
individual fitness vs. group fitness, 157–158
Fixation
definition/description, 246
driving genes/alleles, 96, 97
probabilities of, 43
Forced relocations
examples, 145–146
gene flow, 145–146
Fox domestication, 6, 112
Freudian theory, 190

G6PD variants and malaria defenses, 73, 88, 89, 143, 217
Game-theory analysis, 72
Gardner, Howard, 208
Gateway mutations, 23
Gaucher's disease
Ashkenazi Jews, 188, 213, 214, 221 (fig.), 246
definition/description, 246
Gell-Mann, Murray, 191
Gene definition/description, 246
Gene flow
barriers to, 138–141
blue eyes, 18, 92, 148–153, 150 (photo), 151 (photo)
colonization, 142–144
deserts' effects, 139
effect of oceans, 138–139
forced relocations, 145–146
historical patterns, 141–142
military movements, 144–145
Roman Empire fall and, 148–153
Sarmatians, 147, 148
slavery, 139, 153
time and, 145
trade, 141–142
*See also* Selective sweeps
Genetic history studies
advantageous alleles, 130–131
mitochondrial DNA use, 129–130
overview, 129–131
*See also* Y chromosome
Genetic isolation/Ashkenazi Jews
intermarriage prohibition, 107–108, 193–194, 204, 205, 206, 219–220, 224
persecution, 195, 196, 197–198, 201, 212
Genghis Khan, 106, 145, 184
Genotype, 246
Gibbon, Edward, 152
Gimbutas, Marija, 179
Giovaneli, Rosa, 134
Goddard, Henry, 211
Goleman, Daniel, 208
Gould, Stephen Jay
evolution speed, 5
human evolution, 1
IQ testing, 211–212

Government
  aggressiveness trait and, 111–112
  development, 105
  elites, 109–110, 111–112
  hunter-gatherers and, 105, 111, 113
  standard of living and, 110
  taming/domesticating people, 111–112, 113
Great Danes, 7 (photo), 13, 15, 16
Great Revolt (AD 65–73), 204
Greeks
  as colonizers, 142–143, 145
  kingdoms in Afghanistan/Pakistan, 145
  Pathans and, 145
Green Revolution, 111
Grotte des Fées (Fairy Grotto), 29
Group fitness vs. individual fitness, 157–158
Group selection, 246–247
Gunpowder invention/effects, 184
*Guns, Germs, and Steel* (Diamond), 66, 121

Hair color, 94, 112
Haldane, J. B. S., 42
Hamilton, William, 118
Hanno the Navigator, 170
Haploid organisms, 247
Haplotype
  definition/description, 20–21, 247
  length meaning, 21, 22
  recombination, 21
HapMap
  definition/description, 20, 247
  lactose tolerance, 84
  skin color, 91
Haroun al-Rashid, 195
Hawk-dove game, 13, 72–73
Hazara, 145
HDL cholesterol, 133, 134
Head Start, 210
Height, normal distribution, 124
*Helicobacter pylori*, 90
Hemochromatosis, 165
Heritability, 247
Herodotus, 144, 170, 176, 182
Heterozygote, 247
Heterozygote advantage
  definition/description, 72, 216, 217
  malaria defenses, 72, 216–217
  whippets/myostatin mutation, 217–218, 218 (photo)
High altitude adaptations, 56
Himalayas and gene flow, 139–140
Hittites/Hittite Empire, 109, 177, 178
HIV/AIDS, 46, 47 (fig.)
HLA system
  Amerindians, 160–161
  functions, 160
  variability, 160–161
Hobsbawm, Eric, 202
Holocene
  agriculture developments, 31
  definition/description, 246, 247
  human/domesticated animal changes, 112
*Homo erectus*, 244, 247
*Homo heidelbergensis*, 44, 247
*Homo neanderthalensis*, 54. *See also* Neanderthals
*Homo sapiens*, 54, 247. *See also* Modern humans
Homozygote, 247
Honey bees
  climate adaptation, 50–51
  introgression, 50–52
  origins/expansions, 50
Hoodbhoy, Pervez, 127

Human genome, 20
Human personality variations
  farmers vs. hunter-gatherers, 113–118, 119
  genetic basis, 56
  *See also* Social patterns, modern
Human population differences
  as genetic differences (overview), 90–91
  genetics of skeletal change, 95
  as "superficial," 90
  *See also specific differences*
"Human revolution," 32. *See also* Upper Paleolithic innovations
Hunter-gatherers
  diet, 76
  famine and, 81–82
  high paternal investment, 116
  infectious diseases, 87
  marital distance, 136
  personality traits, 113–118, 119
  property and, 114–115
  violence/population size, 81
  *See also specific groups*

Ice Age, 8–9, 27, 67
*Iliad*, 19, 177, 183
Imhotep, 1
Incan Empire, 163
Individual fitness vs. group fitness, 157–158
Indo-Aryans, 185
Indo-Europeans
  definition/description, 247
  languages, 174–175, 177, 247
  *See also* Proto-Indo-Europeans
Indus civilization
  definition/description, 247–248
  Indo-Aryans, 185
  studying, 120
  trade, 141–142
Infant mortality, 76
Inner ear, 99–100
Intelligence
  environment and, 210
  as heritable, 209–210, 223
  types of, 208–209
  *See also* Ashkenazi Jews/intelligence; IQ
Interest rates, 116–117
Interfertility, 37
Introgression
  cows, 48, 49 (photo)
  definition/description, 248
  examples, 48, 50–52
  honeybees, 50–52
  invisible vs. visible effects, 52
  modern human/Neanderthal matings, 36, 40, 46, 52, 57, 61–62, 63, 64
  plants, 48
  wild populations, 48, 50
IQ
  Ashkenazi Jews, 188, 191, 210–213, 222, 223–224
  measurements/measuring, 188, 208, 209
  as performance predictor, 188, 189, 208, 209
  *See also* Intelligence
Islamic world
  Jews and, 194, 200–201
  lack of science, 127
Isotope, 248

Jacobi, Carl, 202
Jamestown settlement, 163, 164
Javelins, 32
Jewish communities
  Ashkenazi Jews comparison, 212–213, 224
  in the Islamic world, 194, 200–201
  overview, 187
  *See also* Ashkenazi Jews

Jewish Diaspora, 193, 194, 198–199
Jewish quotas, 212
Jones, Sir William, 175
Justinian, Emperor, 152

Kalonymus family, 195
Kamin, Leon, 211
Kerr, Warwick, 50–51
Klein, Richard, 29–30, 32
Knives, 29
Kronecker, Leopold, 202
Kurds, 150
Kurgan burials, 183
Kurgan hypothesis, 179

Lactase
- continuing production, 77–78
- definition/description, 248
- stopping production, 22, 77

Lactose, 248
Lactose tolerance
- Africa, 77–78, 139, 185
- Arabian peninsula expansion, 185
- description, 12
- increase/evidence, 22, 83–84
- Proto-Indo-European expansion, 174, 181–186
- sweep, 22, 77–78, 83–84, 132, 139, 174, 181–186

Language capabilities
- FOXP2 gene, 63
- information transmission across generations, 27
- limits without, 27
- modern humans, 26–27, 45, 156
- new genes affecting, 99–100

Lascaux cave painting, 35 (photo)
Lemme, Chuck, 14
Leprosy, 161
Lewontin, Richard, 15, 17
Limone sul Garda villagers, 133–135, 133 (photo), 227
Linkage disequilibrium
- definition/description, 61, 91–92, 248
- modern human/Neanderthal matings, 61
- skin color change, 91–92

Lion Man of Hohlenstein, 39 (photo)
Livingston, David, and wife, 172
Locus, 248
Loss of function
- definition/description, 248

Lost Dutchman Mine, location, 400
Lymphatic filariasis, 161–162

Maimonides, 201
Malaria
- falciparum type, 16, 73, 87, 89, 104, 109, 162, 171, 172
- locations, 87, 139, 143, 159
- sub-Saharan Africa
  - colonization attempts, 139, 171–172, 173

Malaria defenses
- alleles becoming common, 44
- alpha-thalassemia, 88, 171, 217
- beta-thalassemia, 88, 143, 217
- concentration in few metabolic pathways, 217
- Duchenne muscular dystrophy, 88
- G6PD variants, 73, 88, 89, 143, 217
- Greeks/Phoenicians and, 143–144
- Hemoglobin E, 55
- hemoglobin molecule, 13, 55, 73, 88, 89, 217

heterozygote advantage, 72, 216–217
overview, 87–89
as recent, 89
sickle cell, 16, 55, 72, 88, 216, 217
side effects, 16, 55, 72, 73, 87–89, 216, 217
strategy differences, 55
Malthus, Thomas, 100, 102
Malthusian societies, 70, 81
Malthusian trap
agriculture transition, 100–102, 103, 104–105, 117, 118
definition/description, 248
disease, 100, 103, 104
famine/malnutrition, 100, 102, 103, 104
overview, 100–105
war, 102, 103
*Man with the Hoe* (Millet), 110 (fig.)
Mandan Indians, 167
Manhattan Project, 190
Manic-depression, 126
Maxwell, James Clerk, 125
Mayr, Ernst, 1
Measles
agriculture/pre-agriculture, 86
Columbian explosion, 161
critical community size, 86, 87
Medicine (Europe 1600s), 168–169
Meiosis, 96
Melanin, 12, 92, 94. *See also* Skin color change
Merovingian monarchs, 195
Mesopotamian trade, 141–142
Mexicans
OCA2 gene, 93
paternal/maternal ancestry, 46
Microcephalin (MCPH1) gene, 62–63
Miele, Frank, 14
Milk, 77. *See also* Lactase; Lactose
Miscarriage rates in humans, 98
Mitochondrial DNA
Africanized bees, 51
Amerindian genetic history, 46, 129–130
definition/description, 248
European bees, 51
modern human/Neanderthal matings, 45–46, 61
Mitosis, 97
Mizrahi Jews, 187, 213
Moctezuma II, 162
Modern human/Neanderthal matings
adaptations to climate, 54
adaptations to European conditions, 53–54
advantageous alleles, 41–42, 43–44, 53, 59, 62, 64
agriculture solutions, 59
alternative strategies development, 56–57, 59
bestiality, 37, 40
controversy, 36–37
disease resistance, 54
genetic evidence, 61–63
how/where, 44–46
interfertility, 37
introgression, 36, 40, 46, 52, 57, 61–62, 63, 64
linkage disequilibrium, 61
natural selection, 57
neutral alleles, 40–41, 42–43, 53
new alleles, 40–44
origin models, 36
skeletal evidence, 59–60
Upper Paleolithic innovations and, 35–37

Modern humans
  in Africa, 54–55
  arrival in Europe, 25
  description/anatomy, 26, 27, 59–60
  diet, 27, 33
  hearing changes, 4–5
  language capabilities, 26–27, 45, 156
  population growth, 65, 66
  tools/weapons, 25, 26, 29, 30, 156, 244
  *See also* Châtelperronian tradition; Upper Paleolithic innovations
Modern humans' displacement of Neanderthals
  alliances, 27
  biological advantages theory, 156
  body build, 26, 27
  "cootie" theory, 28
  diet, 27
  intelligence, 26, 29
  interaction/evidence, 27, 28–29, 45
  language capabilities, 26–27, 45
  racism criticism and, 28
  time scale of, 25–26, 27
  tools/weapons, 25, 26, 29, 30, 156
  trade, 27
Mongol expansion, 155, 156
Moose/deer and parasites, 28
Mousterian industry
  area of, 29
  definition/description, 29, 248
  knives, 29
MtDNA. *See* Mitochondrial DNA
Multiple sclerosis, 161
Muscle
  bursts of strength vs. endurance, 117
  trade-offs with brain development, 99
  whippets/myostatin mutation, 217–218, 218 (photo)
Mutations
  agriculture transition, 59, 66, 74–76
  cholesterol example, 133–135
  definition/description, 74, 248
  driving genes/alleles, 95–98
  gateway mutations, 23
  Limone sul Garda villagers, 133–135
  loss-of-function mutations, 248
  negative vs. positive effects, 132–133
  ongoing sweeping alleles functions, 75–76, 132
  rate/population size, 65–66, 74, 75, 80–81
  time to spread, 65, 66, 74–75
  *See also specific examples*
*Myth of the Jewish Race, The* (Raphael/Jennifer Patai), 203

Napoleon, 202
*Narrow Roads of Gene Land* (Hamilton), 118
Natural selection
  Ashkenazi Jews, 216–222
  camel/transportation analogy, 57–58
  culture and, 3–4
  definition/description, 249
  differentiating human populations, 10, 55, 89–90
  environmental stability, 2–3
  genetic isolation, 218–220
  inferior choices, 57, 58

mechanism with Ashkenazi Jews/genetic diseases, 222–224
as short-sighted, 57, 58
solution diversity, 57, 58
*See also specific examples*
Navajo
diabetes, 80
OCA2 gene, 93
Nazis, 203
Neanderthals
brain development, 54, 55
burials, 34
definition/description, 249
description/anatomy, 26, 54, 55, 59–60
last of, 25–26
locations, 45, 249
views on, 36, 46, 53
way of life, 54, 55
*See also* Châtelperronian tradition; Modern human/Neanderthal matings; Modern humans displacement of Neanderthals; Mousterian industry
Necho, Pharaoh, 170
Neel, James, 81
Neolithic period, 249
Neoteny, 11, 113
Neumann, John von, 190
Neuron, 249
Neurotransmitters, 98
New Guinea highlanders, 140–141
Newton, Isaac, 125
Niall of the Nine Hostages, 106
Niemann-Pick disease, 213, 214, 220
Nobel science prizes, 190
"Noche Triste," 162
Normal distribution (bell curve), 124
Nucleotide, 249

Oak trees in England, 137–138
Occipital bun, 60
Old Stone Age, 30, 249. *See also* Paleolithic period; Upper Paleolithic innovations
Onchocerciasis (river blindness), 161–162

Paleolithic period, 249. *See also* Upper Paleolithic innovations
Parasites, 28
Park, Mungo, 171
Park, Thomas, 171–172
Parsis, 249
Patai, Raphael, and Jennifer, 203
Pathans
definition/description, 145, 249
Greeks and, 145
Pequot War (1636), 113
Perelman, Grigori, 191
Phase transitions, 123, 124–125
Phoenicians, 142–143
Pilgrims, 163, 164
Pirates, 151, 153
Pit bull terriers, 11
Pizarro, Francisco, 163, 169
Placenta, 47
Plagnol, V., 61
Plants. *See* Agriculture transition; Domestication of plants
Pleistocene epoch, 249
Poetry
poets' fitness, 126
Proto-Indo-Europeans, 176–177
Poincaré conjecture, 191

Polish-Lithuanian Commonwealth
 Ashkenazi Jews, 198–200, 202
 science/technology, 198, 202
Polynesians
 diabetes, 80–81
 infectious diseases, 90–91, 169
Pomaroli, Giovanni, 134
Population size control
 birth control, 102, 109
 silphium, 109
Population size/growth
 agriculture transition, 66, 69, 74, 226
 birth control, 102, 109
 elites' reproductive advantage, 104–105, 106–107
 genetic innovations, 66
 innovations, 66
 Malthusian trap, 100–105
 mutations, 65, 66
 statistics, 65, 69
Positive selection, 249
Primates, 249
Property
 agriculture transition, 70, 114–115, 116–117
 hunter-gatherers, 114–115
Protein
 agriculture transition/diet changes, 76
 definition/description, 249
Proto-Indo-European expansion
 dairying advantages, 181, 184–185
 domestication of horse/mounted warriors, 176, 179, 180
 homeland (Urheimat) location controversy, 177–179
 Kurgan hypothesis, 179
 lactose tolerance mutation, 174, 181–186
 mobility advantage, 182–183
 Renfrew's model, 178–179
Proto-Indo-Europeans
 epic poetry, 176–177
 as grain farmers, 175–176, 180
 metallurgy, 176, 177
 religion, 176
 social classes, 176
 social system, 176
 as stock raisers, 175–176, 180
 as warriors, 176, 183–184
 *See also* Indo-European
Puritans, 163
Pygmies, 62, 170
Pyramidal neuron, 249

Rabbinical Judaism, 193, 194
Rabbits in Australia, 131–132
Recessive allele. *See under* Alleles
Recombination
 definition/description, 21, 62, 250
 microcephalin gene, 62
Renfrew, Colin, 178
Rheumatoid arthritis, 162
*Rig Veda*, 177
Rinderpest, 48
River blindness (onchocerciasis), 161–162
RNA, 46
RNA viruses, 46–47, 47 (fig.)
Roman Empire fall, 148–153
Rousseau, Jean-Jacques, 105

Sahara Desert, 139, 170–171
Salt conservation gene, 71–72
Sargon II, 146
Sargon of Akkad, 1, 141
Sarich, Vince, 14

Sarmatians
beliefs/legends, 148
definition/description, 146–148, 250
gene flow, 147, 148
language, 250
Scythians and, 250
Schistosomiasis, 161–162
Schwinger, Julian, 190–191
Science
beginnings, 125
connectivity effects, 125–126, 127
factors important to, 126–128
genetics and, 122–128
locations without, 127
protoscience, 125
Western Europe vs. Polish-Lithuanian Commonwealth, 198, 202
Sculpture, 34, 38–39 (photo)
Scythians
definition/description, 250
Persian Empire invasion, 182–183
Sarmatians and, 250
social system, 176
way of life, 182–183
Selective sweeps
agriculture, 157
ApoA-I protein variant example, 133–135
definition/description, 250
driving genes/alleles, 96–97
formula for spread/application, 136, 137
isolated populations and, 135
ongoing sweeps, 75
process beginning, 132–133
sweeping alleles
categories/functions, 75–76, 132
time to become common, 134–135
village-to-village contact, 135–137, 138
"well mixed" populations, 132, 135
Sennacherib, 146
Sephardic Jews, 187
Sepsis risk, 12–13
Serotonin, 98, 112–113
Shingles, 86
Silphium, 109
Skeletal record
brow ridges, 94, 95
change overview, 94–95
milk drinking effects, 183
modern human/Neanderthal matings, 59–60
*See also* Anatomy; Brain development; Teeth
Skeptics Society, 14
Skin color and vitamin D, 78, 92, 93–94, 243
Skin color change
albinism, 92–93
drivers in Europe vs. Asia, 72, 91, 93–94
linkage disequilibrium, 91–92
OCA2 allele, 92–93
rapidity of, 18, 92
as recent, 91–92
skin cancer, 44
and Vitamin D/diet, 78
Slash-and-burn agriculture, 87
Slavery
Africa, 139, 141, 153, 171
Ashkenazi Jews, 204–205

Slavery *(continued)*
gene flow, 139, 153
"tameness" and, 113
Sleeping sickness, 172
Smallpox
Amerindian fatality rate, 167
Columbian explosion, 161, 162, 165–166, 167, 169
European defenses against, 165, 166
European fatality rate, 167
Mandan Indians, 167
SNPs definition/description, 61, 250
"Social intelligence," 208–209
Social patterns, modern
advantageous alleles, 122–123
age of transition to agriculture effects, 118–119, 120, 121–128
cheaters, 56, 120
culture vs. genetic change, 121–122
distribution of personality traits, 119
high-trust society, 119–120
mix of personality types, 120
outliers, 123–125
overview, 119–120
personality traits of farmers/hunter-gatherers, 113–118, 119
phase transitions, 123, 124–125
populations with slow modernization, 122
science/technology adaptation, 122–128
studying, 120
thresholds, 123, 124
trade, 118–119
Spanish explorers, 46, 129–130
Spanish fly
definition/description, 250
European doctors use of, 168
Sperm production
description, 96
driving genes/alleles, 96, 98
SPAG6 sperm motility gene, 96
Sphingolipids
definition/description, 250
storage disorders, 214, 215, 220
Squanto, 163
Squirrels and viral disease, 28
Stomach cancer, 90
Stomach ulcers, 90
Stoza, 152
Sub-Saharan Africa
African advantages, 170–173
agriculture/adaptations, 80
Dutch colonization, 172
European advantages, 172, 173
expansion into, 139, 170–173
FAS, 83
gene flow, 139
lack of science, 127
OCA2 albinism, 93
technology comparison, 172
Suebians, 152
Sumerians, 180
Sungir, 33
Surui people, 167
Sweeps. *See* Selective sweeps
Syncytin, 47
Syphilis, 162

Taino people, 163
Talmud, 193, 202
Taurine cattle, 48, 49 (photo), 68
Tay-Sachs disease, 188, 203, 213, 214, 220, 222, 250
Taylor, Elizabeth, 203

Teeth
changes with domestication, 112
evolutionary changes, 4, 60, 112
high-carbohydrate diet effects, 79
iron deficiency effects, 77
retromolar space changes, 60
Teosinte, 6, 8 (fig.), 13
Texas longhorn, 49 (photo)
*Third Chimpanzee, The* (Diamond), 140
"Thrifty genotype" hypothesis, 81
Thucydides, 227
Tibetans and high altitude, 56
Tiglath Pileser III, 145–146
Titus' destruction of the Temple, 193
Tolstoy, Leo 186
Tooby, John, 9–10
Tools/weapons
Aurignacian culture, 26, 29, 244
bone/ivory use, 32
Bushmen, 3–4
Châtelperronian tradition, 29, 156, 245
iron tools of Bantu, 156
modern humans, 25, 26, 29, 30, 156, 244
Mousterian industry, 29, 248
Paleolithic age, 249
projectile weapons, 26, 32–33, 156
regional style beginnings, 30
sewing needles, 25
trade for materials, 30
Upper Paleolithic innovations, 30, 32–33
*See also specific tools/weapons*
Torah, 193, 201
Torsion dystonia, 221–222
Trade
Ashkenazi Jews, 195–196
farmer personality traits and, 118–119
gene flow, 141–142
infectious diseases, 87
middlemen/agriculture, 119
modern humans' displacement of Neanderthals, 27
modern social patterns, 118–119
overview, 141–142
tools' materials, 27, 30
Upper Paleolithic, 30
Triglycerides, 133
Tuaregs, 150, 150 (photo)
Tuberculosis, 166–167
Tungiasis, 162
Turks' expansion, 155, 156
Tutsis, 77–78
Typhus, 86

Upper Paleolithic innovations
art, 30, 33, 34–35, 35 (photo)
burials, 33–34
cave paintings, 30, 34, 35 (photo)
description, 30, 32–35, 225–226
fire uses, 33
food preservation, 33
genetic change, 31–32
hunting, 32–33
overview, 29–32, 64
pottery, 33, 34
protective structures/dwellings, 34
rate, 30
sculpture, 34, 38–39 (photo)
social system, 33
tools/weapons, 30, 32–33

Upper Paleolithic innovations *(continued)*
  trade, 30
  *See also* Modern human/Neanderthal matings

Vandals
  blue eyes, 149, 153
  definition/description, 151–152, 250
  Roman Empire and, 149, 151, 152, 250
Venus of Dolni Vestonice, 34, 38 (photo)
Venus of Willendorf, 34, 38 (photo)
Viruses
  retroviruses, 46–47, 47 (fig.)
  squirrels and, 28
Visigoths, 152
Vitamin D
  agriculture transition/diet changes, 78
  deficiency problems, 78
  skin color, 78, 92, 93–94, 243
  ultraviolet radiation, 78
Vitamin-deficiency diseases, 76

Wall, J. D., 61
Wars
  biological superiority, 158
  "flower wars" of Aztecs, 164
  Malthusian trap, 102, 103
Weapons, 26, 32–33, 156. *See also* Tools/weapons
Weinryb, Bernard D., 199–200
West Indies, 163
Western Europe honey bees, 50, 51
Wheat, 48
Wheel invention/use, 178, 180
Whippets/myostatin mutation, 217–218, 218 (photo)
Whitfield, Charles, 50, 51
Whooping cough, 161
Witten, Edward, 191
Wolves, 7 (photo)
  behavior, 6, 10–11, 13, 55
  deer/moose and, 28
  dogs and, 5, 6, 10–11, 13, 52, 55
Writing invention, 180

X chromosome, 250

Y chromosome
  Amerindian genetic history, 46, 129–130
  definition/description, 251
  elites reproductive advantage evidence, 106
  modern human/Neanderthal matings, 45–46, 61
Yanomamo
  definition/description, 250–251
  tuberculosis, 166–167
Yellow fever, 139, 162, 172

Zayed, Amos, 50, 51
Zebu cattle, 48, 49 (photo), 68